T0075954

Analog Circuits and Signal Processing

Series Editors

Mohammed Ismail, The Ohio State University
Mohamad Sawan, École Polytechnique de Montréal

For further volumes:
http://www.springer.com/series/7381

Cristina Azcona Murillo • Belén Calvo Lopez
Santiago Celma Pueyo

Voltage-to-Frequency Converters

CMOS Design and Implementation

 Springer

Cristina Azcona Murillo
Faculty of Sciences
University of Zaragoza
Zaragoza, Spain

Belén Calvo Lopez
Faculty of Sciences
University of Zaragoza
Zaragoza, Spain

Santiago Celma Pueyo
Faculty of Sciences
University of Zaragoza
Zaragoza, Spain

ISBN 978-1-4614-6236-1   ISBN 978-1-4614-6237-8 (eBook)
DOI 10.1007/978-1-4614-6237-8
Springer New York Heidelberg Dordrecht London

Library of Congress Control Number: 2012956310

Printed on acid-free paper

Springer is part of Springer Science+Business Media (www.springer.com)

Preface

Wireless sensor networks (WSNs) are an emerging technology and certainly will be, without any doubt, a key technology in the future. To name a few, in 1999, the Business Week magazine included them as one of the most important technologies for the twenty-first century; in 2003, the Technology Review published by the MIT listed them as one of the top ten emerging technologies; and in 2006, Nature magazine dedicated a piece of news to them, entitled "2020 Computing: Everything, Everywhere," where WSN was considered to cause a huge impact in the near future. Now, in 2012, a big progress has been made in the development of these networks, and there is a continuously accelerated advancement of this technology to satisfy the demands of the market.

A WSN consists of a set of spatially distributed sensor nodes that monitor different parameters, process the information using the embedded microcontroller (μC), and send measured data through a suitable RF module. Within the sensory field, a critical element in a WSN node is the electronic interface that conditions the signal coming from the set of low-cost sensors into a signal appropriate to be entered into the μC, task which usually consists in the amplification of the sensor output signal and the elimination of the offset, followed by its analog-to-digital conversion (ADC). Focusing in the ADC function, instead of conventional ADCs, the advantages of quasi-digital converters, such as voltage-to-frequency converters (VFCs), can be fully exploited offering a timely solution to perform signal digitalization.

A VFC is a data converter that codifies the information in frequency: the output signal is equivalent to a serial digital signal without the need of synchronization that can be directly interfaced to the μC using a single digital port, where the final digitalization is made thanks to the internal μC clocks. Thus VFCs combine the simplicity inherent to analog devices with the accuracy and noise immunity proper to digital converters. Besides noise immunity, VFCs usually need less integration area and power consumption compared with classical ADCs of equivalent accuracy and resolution used in smart sensor conditioning. In addition, due to cost reduction, low-cost μCs are used in WSN with the disadvantage of having the number of digital ports very limited, which prevents the use of conventional ADCs.

In this scenario, the main objective of this book is to develop VFC solutions integrated in standard CMOS technology to be used as a part of a microcontroller-based multisensor interface in the environment of portable applications, particularly within a WSN node. To use these converters in battery operated systems, stringent specifications to fulfil are low voltage, to be fed with the same batteries than the portable system and low power, to extend the batteries life. Thus, all through this book we offer a detailed study about VFC design—both at basic VFC cells and system level—their main characteristics and performances.

The starting point is the definition of the main VFC characteristic parameters, followed by a complete review of the most common types of VFCs existing in the literature, the multivibrator and the charge-balance VFC, and a brief analysis of the main frequency-to-code conversion methods, issues addressed in Chap. 2.

Next, as the multivibrator VFC is the one selected to be implanted for its use in a WSN node, its basic blocks, i.e., voltage-to-current converters, bidirectional current integrators, control and bias circuits are deeply examined. These basic blocks must be carefully designed as they will limit the full VFC performance, so their specifications have to guarantee those required by the corresponding application. Thus, the main challenges and solutions encountered during the design of such high performance cells are summarized in Chap. 3 and different high performance integrated proposals that will be next employed in specific VFCs are described and characterized considering low-voltage low-power constraints associated to portable systems. To achieve low-power consumption and easy any future scaled to shorter transistor channel length technologies, low-voltage power supplies have been employed: this requires greater effort in the design, but guarantees the validity of the achieved results in current submicron process technologies.

To close, the work is focused on the complete characterization of few different VFCs making use of some previously studied cells. Four complete VFC proposals are fully designed and evaluated in Chap. 4: a programmable VFC that includes an offset frequency and a sleep/mode enable terminal; a low-power rail-to-rail VFC; and two rail-to-rail differential VFCs. All of them are temperature compensated. These novel VFC contributions are more than competitive with those already presented in the literature and demonstrate their feasibility to be used in WSN interfaces as an alternative to conventional ADCs.

Zaragoza, Spain   Cristina Azcona Murillo
                                                                             Belén Calvo Lopez
                                                                        Santiago Celma Pueyo

Contents

List of Figures

List of Tables

Parameter Glossary

Parameter	Significance
C	Equivalent capacitance of an OTA
Cc	Compensation capacitor
C_{gs}, C_{gd}	Gate-source and gate-drain capacitance
C_{int}	Integrating capacitor
f_0	Output frequency
f_{clk}	Clock frequency
f_x/T_x	Frequency/period of the unknown frequency
G	Equivalent transconductance of an OTA
g_m	MOS transistor transconductance defined as $\delta I_D/\delta V_{GS}$
G_M	Total transconductance of a V-I converter
g_{mb}	MOS transistor bulk-transconductance defined as $\delta I_D/\delta V_{BS}$
I_B	Bias current
I_{DO}	Off-drain current of a MOS transistor
I_{out}	Output current
K	Scaling factor between current mirrors
K_B	Boltzmann constant (1.38×10^{-23} J/K)
L	Channel length of a MOS transistor
N	Number of bits
n	Subthreshold slope factor
N_x	Number of pulses of a given frequency
q	Electron charge (1.602×10^{-19} C)
R^2	Linear regression coefficient (R)
R	Equivalent resistance of an OTA
R_0	Nominal resistance
R_C	Compensation resistor
R_L	Load resistor
r_0	MOS transistor output resistance
R_S	V-I conversion resistor

S	Sensitivity
T	Temperature
TC	Thermal coefficient
T_W	Gate time (frequency-to-code conversion)
V_{cap}	Voltage across integrating the capacitor C_{int}
V_{DD}	Supply voltage
$V_{DS,sat}$	Drain-source saturation voltage of a MOS transistor
V_E	Early voltage
V_{GS}, V_{DS}, V_{BS}	Gate-source, drain-source and bulk-source MOS transistor
V_H	High comparison limit
V_{in}	Input voltage
V_L	Low comparison limit
V_T	Thermal voltage ($V_T = K_B T/q$)
V_{th}	Threshold voltage of a MOS transistor
W	Channel width of a MOS transistor
α	Attenuation factor
ΔV	Difference between the comparison limits ($\Delta V = V_H - V_L$)
λ	Channel-length modulation factor of a MOS transistor
ω	Angular frequency

List of Acronyms

Acronym	Significance
μC	Microcontroller
ADC	Analog-to-Digital Converter
ASIC	Application-Specific Integrated Circuit
BSIM	Berkeley Short-channel IGFET Model
BW	Bandwidth
CFC	Capacitance-to-Frequency Converter
CMOS	Complementary Metal-Oxide-Semiconductor
CMRR	Common-Mode Rejection Ratio
COTS	Commercial Off-The-Shelf
DAC	Digital-to-Analog Converter
DC	Direct Current
DCM	standard Direct Counting Method
dVFC	Differential Voltage-to-Frequency Converter
EEPROM	Electrically Erasable Programmable Read-Only Memory
FBVA	FeedBack Voltage Attenuation
FFCA	FeedForward Current Attenuation
FFVA	FeedForward Voltage Attenuation
FPAA	Field-Programmable Analog Array
FS	Frequency Span
GSM	Global System for Mobile Communications
HRP	High Resistance Polysilicon
IC	Integrated Circuit
ICM	Indirect Counting Method
IFC	Current-to-Frequency Converter
LP	Low-Power
LV	Low-Voltage
MIM	Metal-Insulator-Metal
MUX	Multiplexer

NMOS	Negative-Channel Metal-Oxide-Semiconductor
OA	Operational Amplifier
OTA	Operational Transconductance Amplifier
PCB	Printed Circuit Board
PMOS	Positive-Channel Metal-Oxide-Semiconductor
PND	P^+ non-salicide diffusion
PSRR	Power Supply Rejection Ratio
PVA	Programmable Voltage Adapter
RAM	Random Access Memory
RF	RadioFrequency
RFC	Resistance-to-Frequency Converter
RISC	Reduced Instruction Set Computer
SNR	Signal-to-Noise Ratio
ST	Schmitt Trigger
SVFC	Synchronous Voltage-to-Frequency Converter
THD	Total Harmonic Distortion
UISI	Universal Intelligent Sensor Interface
UMC	United Microelectronics Corporation
UMSI	Universal Micro-Sensor Interface
USB	Universal Serial Bus
USI	Universal Sensor Interface
USIC	Universal Sensor Interface Chip
UTI	Universal Transducer Interface
UV	UltraViolet
VCO	Voltage-Controlled Oscillator
VFC	Voltage-to-Frequency Converter
VWC	Voltage-Window Comparator
WSN	Wireless Sensor Network

Chapter 1
Introduction

At present wireless sensor networks (WSNs) are an emerging technology and certainly will be, without any doubt, a key technology in the future. In 1999, the Business Week magazine included them as one of the most important technologies for the twenty-first century [BUS99]. In 2003, the Technology Review, published by the MIT (Massachusetts Institute of Technology), listed them as one of the top ten emerging technologies [TEC03]. Nature magazine also has dedicated a piece of news to them, entitled "2020 Computing: Everything, Everywhere," where WSN was considered to cause a huge impact in the near future [BUT06]. The ABI (Allied Business Intelligence) Research principal analyst ensured that "in 2010, WSN IC shipment grew more that 300 % over the previous year." Now, in 2012, it can be said that a big progress has been made in the development of these networks, and that there is a continuously accelerated advancement of this technology. Therefore, it is reasonable to expect that in 5–10 years, the world will be covered with these networks. According to ABI research predictions, the number of nodes that are going to be used into home appliances is forecast to increase from 8.5 million in 2010 to 242 million a year by 2015, as shown in Fig. 1.1 [ABI11].

Sensor networks have been used for many years, but they used to be wired, with power consumption not being a constraint. However, this changed with the development of WSN, technology first motivated by military applications, such as battlefield surveillance [KHE05]. A WSN consists of a set of spatially distributed sensor nodes organized into a cooperative network. These sensor nodes monitor different parameters depending on the target application, usually physical and chemical magnitudes such as temperature, humidity, or pH. Once acquired, they process and linearize the information, and correct cross-related measurement errors using the embedded microcontroller (μC). This information is stored and sent by means of wireless communication protocols to a coordinator, which forwards it to a central node that process the received data to analyze, interpret and act when applicable. Thus, WSN have a wide variety of applications that include not only those that traditionally existed for sensor networks but also a number of innovative applications previously unthinkable because of the need for wires. These new applications include health-care monitoring, military applications, industrial

C.A. Murillo et al., *Voltage-to-Frequency Converters: CMOS Design and Implementation*, Analog Circuits and Signal Processing, DOI 10.1007/978-1-4614-6237-8_1, © Springer Science+Business Media New York 2013

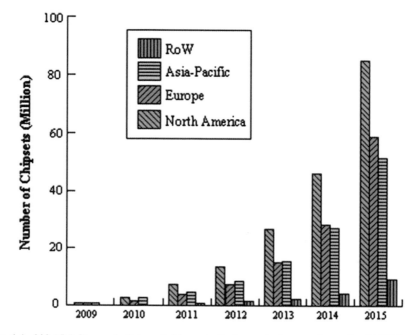

Fig. 1.1 802.15.4. Forecast of chipset shipments for home deployment by region [ABI11]

process control, smart spaces, or environmental applications. In particular, WSN systems are a paradigm in the field of environmental intelligence, where several environmental parameters, such as temperature, UV radiation, and CO_2 concentration, can be acquired with a spatial and temporal resolution almost impossible to achieve with traditional sensing systems [DON08, KRI05, LEE09, LU09, MAH07], paving the way to novel environmental applications, which include early fire forest detection, seismic studies, contaminant transportation control, ecosystem monitoring, or precision agriculture.

The nodes that make up the WSN are also known as motes. Each mote, as shown in Fig. 1.2, is an embedded hardware/software system, which consists of a μC that controls and synchronizes the sensor data acquisition process, the transceiver operation and the memory access and storage; a RF transceiver with a dipole antenna that sends wirelessly the information (they also can include a USB or a GSM module); a power source (batteries, sometimes rechargeable with energy harvesting techniques); and a set of sensors that measure the required parameters. These sensors are followed by an electronic interface that conditions all the sensor output signals and digitize them to be read by the microcontroller [KES05, MEI08]. Therefore, the design, implementation and operation of a WSN demand knowledge from an enormous variety of disciplines such as communications protocols and systems, sensors, sensor interfaces, and data processing.

Focusing in the sensory field, a critical element in a WSN node is the electronic interface that conditions the signal coming from the set of low-cost sensors into a

Fig. 1.2 Scheme
of a WSN node

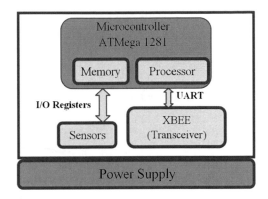

signal, which must be appropriate to be entered into the microcontroller. Usually, the conditioning of a sensor output signal consists in its amplification and elimination of the offset, followed by its analog-to-digital conversion. The interface that performs these tasks has to fulfill some basic conditions: operate with low voltage supply to be battery powered; have low power consumption to extend the battery life; and be able to accommodate different sensors with different output ranges, that is, it must be reconfigurable, programming its operation from the microcontroller and adapting in real time its response depending on the sensor characteristics, thus making easy its application in multi-sensor nodes.

1.1 WSN Sensor Electronic Interface Design

Numerous applications of WSNs involve monitoring physical and chemical parameters over large regions, thus needing a large number of sensor nodes. In these cases, cost reduction becomes a priority, which prevents the use of smart sensors, that is, sensors that include, besides the sensing element, the electronic interface [VAN98], so that the smart sensor provides a standard, digital, and bus compatible output that simplifies the connection to electronic signal processors (μC, computers, etc.). Alternatively, to reduce the cost of these nodes, it is customary to use low-cost analogue sensors along with a programmable electronic interface capable of adapting every sensor output to the port requirements of the μC embedded in the sensing node. Such a reprogrammable sensor interface widens the range of applications and, thus, eases the marketability of the sensing solution. Consequently, in the last decade, the development of sensor interfaces with these features from both, academic and industrial research, has been multiplied.

In a review of the literature, standard commercial sensor interfaces usually include the sensor excitation circuit and a programmable gain amplifier, providing an analog output [ANA12]. There also exist more complex interfaces, which include DACs, ADCs, EEPROMs, and other components that allow compensation of the circuit thermal drifts [MAX01, MAX12]. However, it was not until 1994 when the idea of developing universal sensor interfaces started. These systems are general interfaces that can accommodate different types of sensors with different characteristics. Some of the most popular are the universal sensor interface chip (USIC) [WIL95, WIL96] and the universal transducer interface (UTI) [SMA12]. Other fully integrated universal interfaces are the universal micro-sensor interface (UMSI) [ZHA02] and the universal sensor interface (USI) [CHA04]. Additionally, [MAT10] proposes the universal intelligent sensor interface (UISI), which is based on a commercial available multifunctional chip.

The USIC was the first developed universal sensor interface, within a project carried out by ERA Technology Ltd [WIL95]. It is a complete digital output signal processing chip for data acquisition systems, designed to support a wide range of sensor applications with only a small number of external components [WIL96, ZHA02]. It includes multiplexors that allow routing six sensors, comparators, Σ/Δ demodulators, digital filters, series and parallel interfaces, D/A converters, a RISC processor, and a RAM memory. Among its main characteristics, the high resolution achieved in voltage, current, capacitors, resistors, and frequency measurements has to be highlighted; however, this system was withdrawn from the market because it was too costly [RUI99]. The UTI is simpler because it only includes a complete front-end interface for passive sensors, such as resistive, resistive-bridge, and capacitive sensors. It has a multiplexer that can accommodate six different sensors and a grid of switches that are used for selecting the type of sensor. The resulting output is connected to a relaxation oscillator that generates a period-modulated 3.3 – 5.5 V signal.

These universal interfaces can be useful for a quick prototyping. However, the power consumption of all these interfaces is unsuitable for multisensory battery-operated systems. The USIC and the UMIC have power consumptions around 10 mW, whereas the UTI and the USI demand powers above 7.5 mW, mainly because of the existence of dispensable elements for their use in a WSN sensor interface. Therefore, a custom solution is compulsory.

In such attempt, the conventional solution to provide a sensor output signal that can be read by a digital port of the μC is shown in Fig. 1.3a. It consists of two stages, a programmable voltage adapter (PVA) and an analog-to-digital converter (ADC), preceded by a multiplexer to be able to accommodate the different sensors found in each node. Therefore, first, a sensor is selected, and its output is driven to the PVA. There, an offset is added or subtracted, and the resulting signal is amplified to fit the input voltage of the ADC. Then, the signal is converted into a digital word that is read by the microcontroller. Successive approximation ADCs often are used with microcontrollers because of low cost and ease of interfacing [GAY10, VEN11, ZOU09].

However, for the digitalization of the signal in systems with an embedded microcontroller, the quasi-digital conversion (frequency or time codified information) is becoming a widely used alternative. This is a simple and compact solution that exhibits high immunity to noise and interfering signals [FER07], while takes advantage of the μC time measurement capability [CER10, MEI02, SHE12]. As shown in Fig. 1.3b, the quasi-digital output signal can be directly interfaced to the

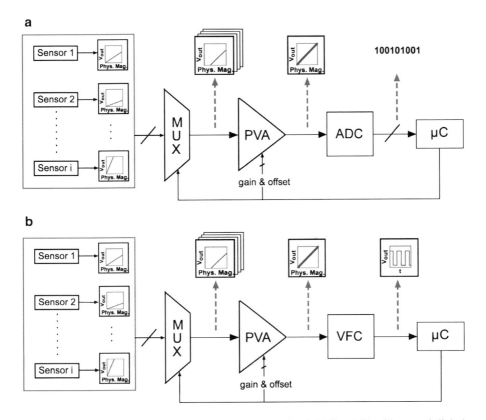

Fig. 1.3 WSN node front-end circuit: (**a**) with a conventional ADC and (**b**) with a quasi-digital converter

microcontroller through a digital port, where the final digitalization is made using the microcontroller internal clocks.

Thus, a quasi-digital conversion made by means of a voltage-to-frequency converter (VFC) is going to be used for implementing the sensor interface because it exhibits some advantages for the same number of bits [MEI08, YUR04]: (1) a VFC is simpler than a conventional ADC; (2) it achieves high noise immunity of the transmission signal, and (3) it achieves high accuracy in the code-to-frequency conversion, with a speed/accuracy compromise that can be mitigated using efficient conversion techniques.

Therefore, traditional analog sensors with VFCs provide a well-timed universal solution for the future microelectronics in this field. The work of this book is placed in this context, trying to offer a reliable alternative to conventional ADC solutions, based on a VFC approach that has not been developed as much as its counterpart. The specifications required for a VFC used within a cost-effective microcontroller-based multi-sensor measurement system include full-range input, low power consumption, supply voltages compatible with batteries, moderate linearity, temperature insensitivity, and output frequency range suitable to be read by the embedded microcontroller.

1.2   Voltage-to-Frequency Converters: State of the Art

As regards the VFC stage, there exist different types of VFCs, the most common are the multivibrator VFC and the charge-balance VFC (asynchronous and synchronous version) that are widely explained in Chap. 2. Thus, a summary of commercial VFCs that can be found in the market is shown in Table 1.1 (multivibrator VFC), Table 1.2 (synchronous charge-balance VFC) and Table 1.3 (asynchronous charge-balance VFC). Note that for each device, only one of its different variants is provided since most of their characteristics are similar.

Table 1.1 Summary of the main performances (max) of commercial multivibrator VFCs

	AD537JH	AD654
Company	Analog Devices	
Single supply: $V_{DD,min}$ (V)	4.5	4.5
Dual supply: $\pm V_{DD,min}$ (V)	± 5	± 5
Quiescent current (mA)	2.5	3
V_{in} (single supply)	0 to $V_{DD} - 4$ V	0 to $V_{DD} - 4$ V
V_{in} (dual supply)	$-V_{DD}$ to $V_{DD} - 4$ V	$-V_{DD}$ to $V_{DD} - 4$ V
Frequency range (kHz)	$0 - 150$	$0 - 500$
Nonlinearity (%)	0.25	0.4
Gain error (%)	± 10	± 10
vs. V_{DD}	± 0.1 %/V	± 0.4 %/V
vs. T	± 150 (ppm/$^\circ$C)	± 50 ppm/$^\circ$C[typ]
T range ($^\circ$C)	0, +70	-40, +85

Table 1.2 Summary of the main performances (max) of commercial synchronous charge-balance VFCs

	AD652JP	AD7741	AD7740
Company	Analog Devices		
Single supply: $V_{DD,min}$ (V)	12	4.75	3
Dual supply: $\pm V_{DD,min}$ (V)	± 6	–	–
Quiescent current (mA)	15	8	1.25
Power down (μA)	–	35	100
Input	By R_{in}	$0 - V_{DD}$	$0 - V_{DD}$
Input current (μA)	500	–	–
R_{in} (kΩ)	20	–	–
Frequency range	$0 - 4$ MHz	$0 - 2.75$ MHz	$0.1 - 1.0$ MHz
Clock frequency (MHz)	5	6.144	1
Nonlinearity (%)	± 0.05	± 0.024	± 0.018
Gain error (%)	± 1.5	± 1.6	± 0.7
vs. V_{DD}	0.01 %/V	0.07 %/V	0.055 %/V
vs. T	± 75 ppm/$^\circ$C	± 16 ppm/$^\circ$C[typ]	± 4 ppm/$^\circ$C[typ]
T range ($^\circ$C)	-55, +125	-40, +105	-40, +105

Table 1.3 Summary of the main performances (max) of commercial asynchronous charge-balance VFCs

	VFC110	ADVFC32k	TC9400	NJM4151	KA331	AD650J	LM231N	VFC32KP	VFC320BP
Company	Burr Brown	Analog Devices	Microchip	JRC	Fairchild	Analog Devices	Texas instruments	Burr Brown	Burr Brown
Single supply: $V_{DD,min}$ (V)	–	–	–	8	5	–	4	–	–
Dual supply: $\pm V_{DD,min}$ (V)	±8	±9	±4	–	–	±9	–	±11	±13
Quiescent current (mA)	16	8	6	6	6	8	4	8	7.5
Input	By R_{in}	By R_{in}	By R_{in}	0, V_{DD}	By R_{in}	By R_{in}	By R_{in}	By R_{in}	750 μA
Input current (μA)	500	250	10	–	156	600	144	250	By R_{in}
R_{in}	25 kΩ	External	External	–	External	External	External	External	External
Frequency range	0 – 4 MHz	0 – 0.5 MHz	0 – 100 kHz	0 – 100 kHz	0 – 100 kHz	0 – 1 MHz	0 – 100 kHz	0 – 500 kHz	0 – 1 MHz
Nonlinearity (%)	1[typ]	0.2	0.25	–	0.01	0.1[typ]	0.01	0.2	±0.1[typ]
Gain error %	±5	±5[typ]	±10[typ]	–	–	±10[typ]	–	±5[typ]	±10
vs. V_{DD}	0.1 %/V	±0.015 % FSR%	–	1 %/V	0.1 %/V	±0.015 FSR/V	0.1 %/V	±0.015 % FSR%	±0.015 % FSR%
vs. T	±100 ppm/°C	±75 ppm/°C[typ]	±40 ppm/°C	±100 ppm/°C[typ]	–	±150 ppm/°C	±150 ppm/°C	±75 ppm/°C[typ]	±50 ppm/°C
T range (°C)	–40, +85	0, +70	–40, +85	–40, +85	0, +70	–25, +85	–25, +85	0, +70	–55, +125

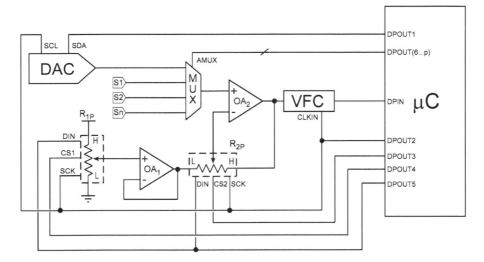

Fig. 1.4 Proposed WSN node front-end circuit with frequency output signal [BAY10]

From these relations of commercial devices, the AD7740 is the only one with supply voltage compatible with standard batteries and with the lowest power consumption. Therefore, a COTS-based custom interface solution targeting low power has been developed using this VFC choice in the research group where this work has been carried out [BAY10], which consists of a simple multiplexed reconfigurable sensor–microcontroller interface meant for low-cost analogue sensors that conditions, digitalizes, and calibrates the signal in a multisensory node (Fig. 1.4). Its performance measurement reveals that the most critical element taking into account the power consumption is the VFC, which means 81 % of the total power consumption during the calibration phase and 99.9 % during the measurement phase (Table 1.4). Two other VFCs formed by commercial devices can be found in [DIM11, WIL05]. Both are fed with $V_{DD} = 5$ V, whereas the supply voltage of a WSN is usually $V_{DD} = 3$ V. In addition, they use discrete capacitors and resistors that directly compromise the good operation of the system. Therefore, the need for a VFC in the form of ASIC is shown because those existing in the market do not fulfill critical requirements, mainly in terms of power consumption.

A review of the literature brings to light also VFCs presented in the form of ASICs, including some converters that are not linear but square-root proportional [JUL08, MAN10]. Focusing on those which are linearly proportional, the majority of them are based on the multivibrator structure that first converts the input voltage into a current; this current next charges and discharges an integrating capacitor between certain voltage limits imposed by a control circuit. From the analysis of their main performances, shown in Table 1.5, they present major disadvantages for their use in autonomous portable applications. Cai and Filanovsky [CAI94] present a supply voltage incompatible with standard batteries. Wang et al. [WAN07] and Yakimov et al. [YAK04] present unsuitable sensitivities to be used with WSN

Table 1.4 Devices used in the proposed interface and their power consumption

Device	Model	Power consumption $(V_{DD} = 3\ V)$	% Consumption in calibration	% Consumption in measuring
DAC	MAX5360[a]	690 µW	18.685	–
MUX	ADG758[b]	0.3 µW	0.008	0.010
OA$_1$, OA$_2$	ISL28194[c]	1.0 µW	0.054	0.067
R$_{1P}$, R$_{2P}$	MAX5400[a]	0.3 µW	0.016	0.020
VFC	AD7740[b]	3 mW	81.237	99.903

[a] Maxim
[b] Analog Devices
[c] Intersil

microcontroller clock frequencies ($f_{clk} = 4$ MHz): in Yakimov et al. [YAK04], the sensitivity is small, having a rather limited effective resolution in the conversion (1.9 kHz/V for an input range of [0.0, 3.0 V]); in the case of [WAN07], the sensitivity is huge (158.84 MHz/V), and therefore, its output cannot be read with typical µC clocks. However, above all these considerations, in terms of input voltage, none of them are rail-to-rail input, a critical issue in modern CMOS processes to optimize the subsequent frequency-to-code conversion resolution. In addition, only Cai and Filanovsky [CAI94] present a temperature compensated design, and, when given, the power consumption makes all these VFCs unsuitable to be used in an interface for low-power applications.

On the other hand, it is interesting that several smart sensors have been recently proposed, with frequency output that converts directly a magnitude into a frequency. They are mainly resistance-to-frequency converters (RFCs) or capacitance-to-frequency converters (CFCs), depending on the sensor to be resistive or capacitive. To mention some, Grassi et al. [GRA06] and De Marcellis et al. [MAR08] present RFCs, that are symmetrically supplied and have a power consumption of 15 and 3.5 mW respectively. Other RFC [MAL05], consumes 30 mW, and Jayaraman and Bhat [JAY07] present a complete ADC based on a resistance-to-frequency conversion, not taking advantage of the use of a microcontroller. Chiang et al. [CHI07] and De Marcellis et al. [MAR09] present CFCs; the first one, with a single supply of 3.2 V, has a power consumption of 6.4 mW, and the second one is made up of commercial devices with symmetrical supply of ±15 V.

1.3   Outline of the Work

The main objective of this book is to develop VFC architectures as part of a low-cost analog-to-digital conversion in a multisensor readout system, suitable to be used in autonomous portable applications such as WSN systems. Therefore, the proposed VFCs: (1) have to be able to handle a rail-to-rail input, which is critical in low-voltage CMOS processes, to enhance the succeeding frequency-to-code conversion (2) the output frequency needs moderate linearity, suitable for driving

Table 1.5 Summary of integrated VFC circuits

	[CAI94]	[YAK04]	[WAN06]	[WAN07]	[CAL09]	[CAL10]
Technology	Bipolar	CMOS FPAA	0.25 μm CMOS	0.25 μm CMOS	0.35 μm CMOS	0.35 μm CMOS
Supply (V)	15	5.0	2.5	2.5	3.0	3.0
Sensitivity	16.67 kHz/V	1.9 kHz/V	520 kHz/V	158.83 MHz/V	1.0 MHz/V	1.0 MHz/V
Input range (V)	0.6 – 6.0	0.0 – 3.0	0.1 – 0.8	0.0 – 0.9	1.0 – 2.0	0.1 – 2.7
Output frequency	0.1 – 100 kHz	1.0 – 6.7 kHz	52 – 416 kHz	0.0 – 55.4 MHz	1.2 – 2.2 MHz	0.1 – 2.7 MHz
Relative error (%)	<2.0	<1.95	<1.0	<8.5	<0.7	<4.0
Linearity error (%)	–	–	–	–	<1	<4
Power (mW)	–	–	–	>1[a]	1.03	0.8
Area (μm^2)	–	–	517 × 596	–	–	440 × 460

[a]Estimated power consumption

low-cost analog sensors, and has to be appropriate to be read by a low-cost μC, and (3) they further have to satisfy the restrictions of LV-LP required for their use in battery-operated systems, such as WSN, without compromising the life of the network. The output frequency needs to be supply voltage independent. This is always desired, but it is more critical in battery powered systems, where the supply voltage is continuously being decreased due to the battery discharging. Another requirement is the temperature independence, to have, for a fixed input voltage, a constant output frequency over all the temperature range. If the VFC is not temperature insensitive, calibration processes have to be carried out in the micro-controller to obtain the input voltage from the output frequency and the temperature, thus increasing the computation time and, as a consequence, the power consumption. A 1.8 V–0.18 μm CMOS process technology is used in this endeavor to profit its main advantages in terms of cost, power consumption and mixed analog-digital integration compatibility: integrating the analog and the digital parts in the same wafer the size is reduced, and the reliability is increased. The proposed VFC structures must meet the following reference specifications: (1) maximum voltage supply $V_{DD} = 1.8$ V, (2) output frequency f_0 below 2 MHz to be successfully read by a microcontroller with $f_{clk} = 4$ MHz, (3) power consumption below 1 mW, and (iv) temperature and supply voltage-independent sensitivity with a linearity error below 0.1 %. These features are good enough for WSN used in environmental applications, where a digitalization of 12 bits can be performed when using the direct counting method measuring during a time window of $T_W = 16.4$ ms.

The starting point has been a theoretical analysis of the main VFC architectures existing in the literature to select the most suitable topology for the target application. Next, the different building blocks forming the architecture of the proposed VFCs, that is, voltage-to-current converters, bidirectional current integrators, control circuits, and bias circuits, will be examined. These basic cells must be carefully chosen because they will limit the whole VFC performances. Thus, the main challenges and solutions encountered through the design of such cells will be summarized, and different high-performance integrated proposals will be described and characterized, considering low-voltage low-power constraints. To facilitate the future technology migration, low-voltage power supply has been used for all of them. The ultimate purpose of this work, focused on the design of complete VFC architectures, includes some features such as digital programmability, rail-to-rail operation, offset frequencies, temperature compensation, or differential processing.

This book is set out in five different chapters; the first one includes this introduction, and the last chapter presents conclusions of the whole work. In all other chapters, a section is reserved at the end for conclusions drawn for that chapter and the employed bibliography. To facilitate the reading of the book, contents, figure and table index, and the list of symbols and acronyms used are offered at the beginning. Finally, two appendixes are added.

This introductory chapter begins by explaining WSNs and their applications, which is the application scenario of this work. Then, the electronic interfaces for multisensory measurement systems are revisited, presenting the advantages of the use of voltage-to-frequency converters instead of classical analog-to-digital converters in these systems. Next, a review of the state of the art of VFCs is carried out, which evinces the necessity of novel low-cost VFC architectures because of the incompatibility in terms of supply voltage, power consumption, and input range of the existing solutions for autonomous portable applications. In the end, the objectives to be achieved are presented, and the book organization is offered.

Chapter 2 covers a theoretical study of the voltage-to-frequency converters. It is divided into three sections. First, the main parameters that describe a VFC, such as input range or linearity, are presented. Then, an analysis of the most common architectures of VFC—the multivibrator VFC, and the asynchronous and synchronous charge-balance VFC—is provided. For each of them, the characteristic block diagram and typical implementations are shown, their main advantages and disadvantages are exhibited, and functional simulations using Cadence® are carried out. Finally, a brief introduction to standard frequency-to-code conversion methods is provided.

In Chap. 3, the study of the main multivibrator VFC cells is completed, as multivibrator is the most suitable VFC to be used in WSN sensor interfaces. The voltage-to-current converter, the bidirectional current integrator, the control circuit, and the bias circuit are separately analyzed, each one in a different section. In the first section, V-I converters are studied, from the conventional scheme to new rail-to-rail solutions, passing through an intermediate approach denominated enhanced V-I converter. The V-I converter compromises the linearity and the input range of the VFC, and largely contributes to the power consumption of the system. Therefore, special attention has to be paid to these features. Because temperature compensation is highly desirable, a common temperature compensation technique is implemented for each of the V-I converters. The proposed V-I converters are characterized, their design is explained, and simulation and experimental results are offered. Then, bidirectional current integrators are explained presenting two different implementation alternatives, both with a grounded capacitor. Next, control circuit operation is introduced, and a proper solution based on a voltage-window comparator is given. High speed comparators are made up with simple differential pairs followed by inverters. A power reduction technique is attained to reduce the power consumption of this block. Finally, bias circuit is analyzed. A bias temperature—and supply voltage—independent current circuit is proposed, and thus, window comparator threshold voltages can be obtained from this reference current.

Chapter 4 offers the implementation of complete VFCs that is accomplished based on the basic cells presented in Chap. 3. Two single-input temperature-compensated VFCs are introduced at this point. The first one is a 1.8 V single-supply, almost full-range VFC, with power consumption below 0.4 mW. It includes programmability, characteristic that adds versatility to the VFC because it can be used with different clock frequencies. The second one is a rail-to-rail input 1.2 V single-supply VFC, with power consumption below 70 µW and an output frequency offset that allows an only frequency-to-code conversion without introducing significant errors in the conversion. Both have been designed in 0.18-µm CMOS technology and have been completely characterized, offering the experimental results.

In addition, two rail-to-rail differential-input VFCs are proposed to process differential signals, for instance, to accommodate Wheatstone bridges. They are designed in 0.18-μm CMOS technology with a single supply of 1.2 V with power consumption below 75 μW; they are temperature compensated, and show immunity to supply voltage variations.

To conclude, Chap. 5 summarizes the general conclusions of the book and the most important contributions. In addition, the research directions that could be further studied in the future are analyzed.

At the end of this work, two appendixes are added. Appendix A summarizes the process features and parameters of the considered 0.18-μm CMOS integrating technology. Appendix B provides a detailed analysis of the frequency response of the voltage-to-current converters presented in this book.

References

[ABI11]   Home automation going mainstream with wireless sensors in consumer appliances. Analyst Insider. http://www.abiresearch.com/press/home-automation-going-main-stream-with-wireless-sen (2011)

[ANA12]   Single-supply sensor interface amplifier. http://www.analog.com-/static/imported-files/data_sheets/AD22050.pdf (2012)

[BAY10]   Bayo, A., Medrano, N., Calvo, B., Celma, S.: A programmable sensor conditioning interface for low-power applications. In: Proceedings of the Eurosensors XXIV Conference, vol. 5, pp. 53–56. Linz, 5–8 Sept 2010 (2010)

[BUS99]   21 ideas for the 21st century. Business Week, pp. 78–167 (1999)

[BUT06]   Butler, D.: 2020 computing: everything, everywhere. Nature **440**(7083), 402–405 (2006)

[CAI94]   Cai, S., Filanovsky, I.M.: High precision voltage-to-frequency converter. In: Proceedings of the 37 IEEE International Midwest Symposium on Circuits and Systems (MWSCAS'94), vol. 2, pp. 1141–1144. University of Southern Louisiana, Layfayette, LA, 3–5 Aug 1994 (1994)

[CAL09]   Calvo, B., Medrano, N., Celma, S., Sanz, M.T.: A low-power high-sensitivity CMOS voltage-to-frequency converter. In: Proceedings of the 52 IEEE International Midwest Symposium on Circuits and Systems (MWSCAS'09), pp. 118–121. Cancun, 2–5 Aug 2009 (2009)

[CAL10]   Calvo, B., Medrano, N., Celma, S.: A full-scale CMOS voltage-to-frequency converter for WSN signal conditioning. In: Proceeding of the IEEE International Symposium on Circuits and Systems (ISCAS'10), pp. 3088–3091. Paris, 30 May to 2 June 2010 (2010)

[CER10]   Ćerimović, S., Keplinger, F., Dalola, S., Ferrari, V., Marioli, D., Kohl, F., Sauter, T.: Smart flow sensor with combined frequency, duty-cycle, and amplitude output. In: Proceedings of the 2010 IEEE Sensors Conference, pp. 580–584. Hilton Waikoloa Village Waikoloa, HI, 1–4 Nov 2010 (2010)

[CHA04]   Chao G., Mejier, G.C.M.: A universal sensor interface chip design in 0.5μm CMOS process. In: Proceedings of the 7th International Conference on Solid-State and Integrated Circuits Technology, vol. 3, pp. 1800–1803. Beijing, 18–21 Oct 2004 (2004)

[CHI07]   Chiang, C.T., Wang, C.S., Huang, Y.C.: A CMOS integrated capacitance-to-frequency converter with digital compensation circuit designed for sensor interface applications. In: Proceedings of the 2007 IEEE Sensors, pp. 954–957. Atlanta, 28–31 Oct 2007 (2007)

[DIM11]   Dimitrov, J.: Inexpensive VFC features good linearity and dynamic range. Electronics
            Design, Strategy, News, pp. 47–48 (2011)
[DON08]   Dondi, D., Bertacchini, A., Brunelli, D., Larcher, L., Benini, L.: Modeling and
            optimization of a solar energy harvester system for self-powered wireless sensor
            networks. IEEE Trans. Industr. Electron. 55(7), 2759–2766 (2008)
[FER07]   Ferrari, V., Ghisla, A., Kovács Vajna, Z., Marioli, D., Taroni, A.: ASIC front-end
            interface with frequency and duty cycle output for resistive-bridge sensors. Sens.
            Actuators A Phys. 138(1), 112–119 (2007)
[GAY10]   Gay, N., Fischer, W.J.: Ultra-low-power RFID-based sensor mote. In: Proceedings of
            the 2010 IEEE Sensors Conference, pp. 1293–1298. Hilton Waikoloa Village
            Waikoloa, HI, 1–4 Nov 2010 (2010)
[GRA06]   Grassi, M., Malcovati, P., Baschirotto, A.: Wide-range integrated gas sensor interface
            based on a resistance-to-number converter technique with the oscillator decoupled
            from the input device. In: Proceedings of the IEEE International Symposium on
            Circuits and Systems (ISCAS'06), pp. 4395–4398. Island of Kos, 21–24 May 2006
            (2006)
[JAY07]   Jayaraman, B., Bhat, N.: High precision 16-bit readout gas sensor interface in 0.13 μm
            CMOS. In: Proceedings of IEEE International Symposium on Circuits and Systems
            (ISCAS'07), pp. 3071–3074. New Orleans, Louisiana, 27–30 May 2007 (2007)
[JUL08]   Julsereewong, A.: Simple square-rooting voltage-to-frequency converter. In:
            Proceedings of IEEE International Conference on Electron Devices and Solid-State
            Circuits (EDSSC'08), pp. 1–4. Special Administrative Region, Hong Kong, 8–10
            Dec 2008 (2008)
[KES05]   Kester, W.: Sensor systems. In: Wilson, J.S. (ed.) Sensor Technology Handbook.
            Elsevier, USA (2005)
[KHE05]   Khemapech, I., Duncan, I., Miller, A.: A survey of wireless sensor networks technol-
            ogy. In: Proceedings of the 6th Annual Post Graduate Symposium on the Conver-
            gence of Telecommunications, Networking and Broadcasting, Liverpool, 27–28 June
            2005
[KRI05]   Krishnamachari, B.: Networking Wireless Sensors. University Press, Cambridge
            (2005)
[LEE09]   Lee, Leibiao, C.H., Lim, L.S.: The optimum configuration of car parking guide
            system based on wireless sensor network. In: Proceedings of the 2009 IEEE Interna-
            tional Symposium on Industrial Electronics (ISIE09), pp. 1199–1202. Seoul, 5–8
            July 2009 (2009)
[LU09]   Lu, B., Gungor, V.C.: Online and remote motor energy monitoring and fault
            diagnostics using wireless sensor networks. IEEE Trans. Industr. Electron. 56(11),
            4651–4659 (2009)
[MAH07]   Mahalik, P.: Sensor Networks and Configuration. Springer, Berlin (2007)
[MAL05]   Malfatti, M., Perenzoni, M., Viarani, N., Simoni, A., Lorenzelli, L., Baschirotto, A.:
            A complete front-end system read-out and temperature control for resistive gas
            sensor array. In: Proceedings of the 2005 European Conference on Circuit Theory
            and Design, (ECCTD'05), vol. 3, pp. 31–34. Cork, 29 Aug to 01 Sept 2005 (2005)
[MAN10]   Maneechukate, T., Yaonun, T., Kamsri, T., Julsereewong, M., Riewruja, V.: Simple
            square-rooting voltage-to-frequency converter using opamps. In: Proceedings of the
            IEEE International Symposium on Circuits and Systems (ISCAS'10), pp. 990–993.
            Paris, 30 May to 2 June 2010 (2010)
[MAR08]   De Marcellis, A., Depari, A., Ferri, G., Flammini, A., Marioli, D., Stornelli, V.,
            Taroni, A.: A CMOS integrable oscillator-based front end for high-dynamic-range
            resistive sensors. IEEE Trans. Instrum. Meas. 57(8), 1596–1604 (2008)
[MAR09]   De Marcellis, A., Di Carlo, C., Ferri, G., Stornelli, V.: A novel general purpose
            current mode oscillating circuit for the read-out of capacitive sensors. In:
            Proceedings of 3rd International Workshop on Advances in Sensors and, Interfaces
            (IWASI'09), pp. 168–172. Trani, 25–26 June 2009 (2009)

[MAT10]   Mattoli, V., Mondini, A., Mazzolai, B., Ferri, G., Dario, P.: A universal intelligent system-on-chip based sensor interface. Sensors **10**(8), 7716–7747 (2010)

[MAX01]   New ICs revolutionize the sensor interface. Sensor Signal Conditioners, Maxim. http://pdfserv.maxim-ic.com/en/an/AN695.pdf (2001)

[MAX12]   Low-cost precision sensor signal conditioner. http://datasheets.maxim-ic.com/en/ds/MAX1452.pdf (2012)

[MEI02]   Meijer, G.C.M., Li, X.: Smart sensor interface electronics. In: Proceedings of the 23rd International Conference on Microelectronics (MIEL'02), pp. 67–74. Niš, 12–15 May 2002 (2002)

[MEI08]   Meijer, G.C.M.: Smart Sensor Systems. Wiley, Chichester, UK (2008)

[RUI99]   Custodio, A., Bragós, R., Pallás, R.: Sensores inteligentes: una historia con futuro. Buran **14**, 13–18 (1999)

[SHE12]   Sheu, M.L., Hsu, W.H., Tsao, L.J.: A capacitance-ratio-modulated current front-end circuit with pulse width modulation output for a capacitive sensor interface. IEEE Trans. Instrum. Meas. **61**(2), 447–455 (2012)

[TEC03]   Culler, D.: 10 emerging technologies that will change the world. Technol. Rev. 106 (1), 33–49 (2003)

[SMA12]   Universal transducer interface (UTI). http://www.smartec-sensors.com/assets/files/pdf/sensors/SMTUTIN.PDF (2012)

[VAN98]   van der Horn, G., Huijsing, J.H.: Integrated Smart Sensors. Design and calibration. Kluwer Academic, The Netherlands (1998)

[VEN11]   de Venuto, D., Stikvoort, E.: Low power smart sensor for accurate temperature measurements. In: Proceedings of the 4th IEEE International Workshop on Advances in Sensors and, Interfaces (IWASI'11), pp. 71–76. Brindisi, 28–29 June 2011 (2011)

[WAN06]   Wang, C.C., Lee, T.J., Li, C.C., Hu, R.: An all-MOS high-linearity voltage-to-frequency converter chip with 520-kHz/V sensitivity. IEEE Trans. Circuits Syst. II **53**(8), 744–747 (2006)

[WAN07]   Wang, C.C., Lee, T.J., Li, C.C., Hu, R.: Voltage to frequency converter with high sensitivity using all-MOS voltage window comparator. Microelectronics J. **38**(2), 197–202 (2007)

[WIL95]   Wilson, P.D., Hopkins, S.P., Spraggs, R.S., Lewis, I., Skarda, V., Goodey, J.: Applications of a universal sensor interface chip (USIC) for intelligent sensor applications. In: Proceedings of the IEE Colloquium on Advances in Sensors, pp. 3/1–3/6. London, 7 Dec 1995 (1995)

[WIL96]   Wilson, P.D., Spraggs, R.S., Hopkins, S.P.: Universal sensor interface chip (USIC): specification and applications outline. Sens. Rev. **16**(1), 18–21 (1996)

[WIL05]   Williams, J.: 1-Hz to 100-MHz VFC features 160-dB dynamic range. Electronics Design, Strategy, News, pp. 82–85 (2005)

[YAK04]   Yakimov, P.I., Manolonov, E.D., Hristov, M.H.: Design and implementation of a V/F converter using FPAA. In: Proceedings of the 27th International Spring Seminar on Electronics Technology Progress, vol. 1, pp. 126–129. Bankya, 13–16 May 2004 (2004)

[YUR04]   Yurish, S.Y.: Sensors and transducers: frequency output vs voltage output. Sens. Transducers Mag. **49**(11), 302–305, 13–16 May 2004, Bankya, Bulgaria (2004)

[ZHA02]   Zhang, J., Zhang, K., Wang, Z., Mason, A.: A universal micro-sensor interface chip with network communication bus and highly-programmable sensor readout. In: Proceedings of the 45th Midwest Symposium on Circuits and Systems (MWSCAS'02), vol. 2, pp. 246–249. Oklahoma State University, 4–7 Aug 2002 (2002)

[ZOU09]   Zou, X., Xu, X., Yao, L., Lian, Y.: A 1-V 450-nW fully integrated programmable biomedical sensor interface chip. IEEE J. Solid-State Circuits **44**(4), 1067–1077 (2009)

Chapter 2
VFC Fundamentals

Voltage-to-frequency converters (VFCs) are, by definition, first-order oscillators whose input is an analog voltage V_{in} and whose output is a frequency signal f_0 linearly proportional to its input voltage, that is, $f_0 = kV_{in}$. They often are denominated as quasi-digital converters because of their analog frequency-coded output. This signal can be directly interfaced to a microcontroller using a single digital input-output port, where, to obtain a digital word, a frequency-to-code conversion must be completed using its internal counters [HUI08]. VFCs are usually mistaken with voltage-controlled oscillators (VCOs), but note that VFCs have different and more stringent performance specifications: typical requirements are high-scale factor accuracy and stability with temperature and supply voltage, wide dynamic range (four decades or more) and low linearity error (less than 0.1 % deviation from zero to full scale) [FRA02, PAL01a].

Although conventional analog-to-digital converters (ADCs) have attracted lots of interest and inversion by huge companies during the past few decades, achieving high levels in terms of development and optimization, currently the widespread use of microcontroller-based measurement systems has made VFCs an attractive alternative to the standard ADC because of their inherent advantages [KIR02a, YUR04]. As mentioned in Chap. 1, frequency modulation exhibits high noise immunity; therefore, frequency signals, compared with analog signals, can be transmitted through communication lines over longer distances [PEA80, SIL00]. Besides, a frequency signal is equivalent to a serial digital signal without the need of synchronization, which makes the interfacing with the microcontroller easier, thus allowing digital conversion with high effective resolution by means of specific conversion algorithms. Note that the frequency-to-code conversion is not a trivial timing window task. Conventional methods have a speed-accuracy trade-off, whereas advanced method needs a license payment [KIR02b, YUR04], which makes the conversion expensive. In addition, it needs a careful analysis that includes the contribution of the algorithmic errors [REV03].

C.A. Murillo et al., *Voltage-to-Frequency Converters: CMOS Design and Implementation*,   17
Analog Circuits and Signal Processing, DOI 10.1007/978-1-4614-6237-8_2,
© Springer Science+Business Media New York 2013

To provide a deep insight into the theory and design of VFC circuits, this chapter reviews the most important parameters that characterize a VFC. Next, the main VFCs types are explained. To close, a brief introduction to standard frequency-to-code conversion methods is provided.

2.1   VFC Characteristic Parameters

This section defines the key parameters that are used to characterize voltage-to-frequency converters [ALL12, ANA12, FRA10, KEN05, PAL01b]. The majority of them are graphically shown in Fig. 2.1, which represents the normalized output frequency f_0 vs. the normalized input voltage V_{in}. Ideal response is shown in black and an actual transfer response is shown in red.

Input range. Set of values between the lower limit $V_{in,min}$ and upper limit $V_{in,max}$, [$V_{in,min}$, $V_{in,max}$], for which the output frequency f_0 varies linearly with the input.

Output range. Set of values between the lower limit $f_{0,min}$ and upper limit $f_{0,max}$, [$f_{0,min}$, $f_{0,max}$] that vary linearly with the given input range.

Frequency span (FS). Positive difference between the output frequencies that correspond to the limits of the input range, where one is the minimum output frequency $f_{0,min}$, and the other is the upper limit of the range $f_{0,max}$. Therefore, $FS = f_{0,max} - f_{0,min}$.

Sensitivity (S) or gain. Variation in the output frequency for a corresponding variation in the input voltage under static conditions. As such, it may be expressed

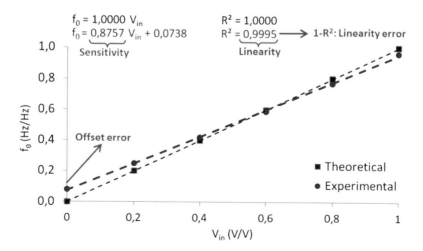

Fig. 2.1 Example of normalized response of an ideal and an actual VFC

as the derivative of the transfer function with respect to the input signal, df_0/dV_{in}. Therefore, as the VFCs considered here are linear, the sensitivity is the slope of the straight line that fits the output frequency f_0 to the input voltage, $f_0 = S\,V_{in}$, given in Hz/V.

Sensitivity or gain error. Deviation of the experimental sensitivity referred to the ideal one, computed as $(S_{exp} - S_{teo})/S_{teo}$ and expressed in percentage.

Sensitivity or gain error drift. Variation of the sensitivity error over the specified temperature range while the remaining parameters are kept constant. It is expressed in ppm/°C.

Relative error. Deviation of the experimental output frequency in each point referred to its ideal value, computed as $(f_{0,exp} - f_{0,teo})/f_{0,teo}$ and expressed in percentage.

Linearity error. Maximum deviation from a straight line passing through the experimental VFC points. There are several measures of this error, the most common one is computed as $(1-R^2)$, R^2 being the linear regression coefficient. It is expressed in percentage or in bits.

Linearity error drift. Variation of the linearity error over the specified temperature range, while the remaining parameters are kept constant. It is expressed in %/°C.

Offset error. Constant frequency added to the output frequency for all the output ranges. It is usually computed as the absolute deviation of the minimum output frequency $(f_{0,exp}(V_{in,min}) - f_{0,teo}(V_{in,min}))$, expressed in Hz or in percentage.

Offset error drift. Variation of the offset error over the specified temperature range while the remaining parameters are kept constant. It is expressed in Hz/°C.

Power supply rejection ratio (PSRR). It is a measure of the VFC dependence with power supply variations, computed as $S^{-1}\Delta f_0/\Delta V_{DD}$ for a constant input voltage, and expressed in dB.

As with most precision circuitry, through adequate calibration processes gain and offset errors can be trimmed by the user in the microcontroller. However, this does not happen with the linearity error, which is inherent to each VFC topology and will determine, with other factors, the accuracy in the final digitalization. In a VFC response where the sensitivity and the offset errors have been trimmed, linearity error remains the same as it was before trimming; therefore, with an ideal sensitivity and an ideal offset, the linearity error shows how well the experimental data fit the theoretical linear response and, therefore, the accuracy of the converter. Thus, it can be considered that for trimmed VFCs, the linearity error is the fundamental parameter, and the smaller the linearity error, the better the VFC, whereas for non-trimmed systems, other errors become important [BRY97, PAL01a].

2.2 VFC Configurations

There are many different approaches for VFCs in the literature, most of them based on the same operation principle, which consists of an alternate integration of the input voltage and the generation of pulses when the integrator output voltage equals a reference voltage [KIR02a]. Focusing on the most recent VCFs, there are two common architectures: the multivibrator VFC and the charge-balance VFC. Their differences can be seen as a different role of the control circuit: in the multivibrator approach the control circuit imposes the threshold voltages, setting the voltage swing in the capacitor, whereas in the charge-balance approach, the control circuit fixes the duration of the charging (or discharging) phase [VID05].

The multivibrator VFC is usually a current-to-frequency converter (IFC), preceded by a voltage-to-current (V-I) converter. It is simple and demands low power; however, it is less accurate than the charge-balance VFC [STO05].

The charge-balance VFC can be either synchronous or asynchronous. It is more accurate than the multivibrator VFC but needs more power. Its output is not a square signal but a pulse train [BRY97].

Next, these three structures—their ways of operation and their advantages and disadvantages—are described. To verify their operation, functional simulations have been carried out. From these simulations the VFC output signal can be observed, as well as the other main signals that characterize the VFC behavior. More detailed descriptions and other VFC implementations can be found at [KIR02a].

2.2.1 Multivibrator VFC

The block diagram of a typical multivibrator VFC is shown in Fig. 2.2a, and its typical implementation is shown in Fig. 2.2b. A multivibrator VFC consists of an input voltage-to-current converter, followed by a bidirectional current integrator driven by a control circuit. The principle of operation is as follows. First of all, the input voltage V_{in} is linearly converted into a current I_{in}. This current alternatively charges and discharges an integrating capacitor C_{int} between two stable voltage limits, V_H and V_L, determined by a control circuit, which is usually a voltage window comparator (VWC) or a Schmitt trigger. The waveform at the integrator output is a linear triangular wave and the output of the control circuit is a square signal (Fig. 2.2c), which is the output frequency f_0 of the VFC and also the signal that controls the direction of the integrating current.

Thus, a fully symmetrical repeated loop is built, obtaining at the output a square signal with a frequency given by

$$f_0 = \frac{1}{2} I_{in} \frac{1}{C_{int} \Delta V} \qquad (2.1)$$

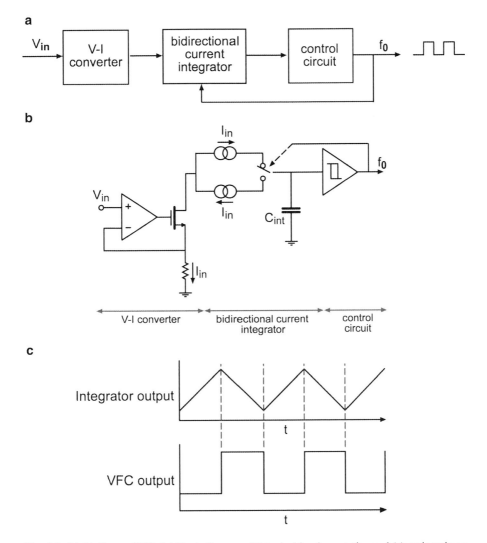

Fig. 2.2 Multivibrator VFC: (**a**) block diagram, (**b**) typical implementation and (**c**) main voltage signals

where I_{in} is the generated current, which is proportional to the input voltage, C_{int} is the integrating capacitor, and $\Delta V = V_H - V_L$.

Practical multivibrator VFCs can achieve around 14 bits of linearity and comparable stability (variation of the actual frequency over time with respect to the ideal frequency), although they may be used in ADCs with higher resolutions without missing codes [ZUM08]. According to (2.1), the performance limits are mainly set by the V_{in}–I_{in} linearity conversion, the comparator threshold noise, and the capacitor temperature coefficient. At high output frequencies, for period widths comparable to VWC delay time, non-negligible errors are introduced in the response of the VFC [KES09].

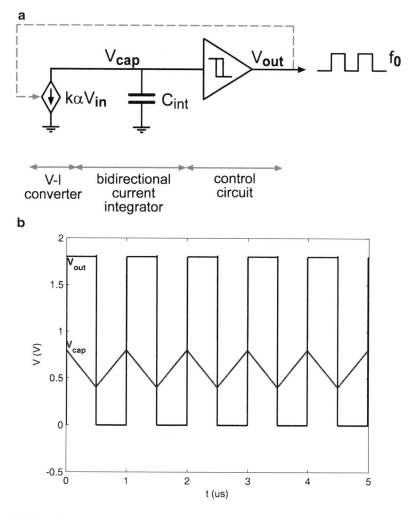

Fig. 2.3 Functional simulation of the multivibrator VFC: (**a**) Schematics and (**b**) capacitor and output waveforms

The scheme shown in Fig. 2.3a was implemented to model the multivibrator VFC in Fig. 2.2a. The input voltage V_{in} is converted into a current $I_{in} = k\alpha V_{in}$ with an ideal voltage-controlled current source. This current is integrated in a grounded capacitor C_{int} between the limits V_L and V_H of a VWC, made up of two ideal comparators and an output $\overline{RS}$ flip-flop to generate stable signals. The output signal of the VWC, that is, the output signal of the VFC, also controls the sign of the current in the bidirectional current integrator. When the output of the VFC is high, $\alpha = +1$, starting the capacitor discharge phase until V_{cap} reaches V_L. At that time, the output of the VFC goes down, $\alpha = -1$, starting the capacitor charge until V_{cap} reaches V_H, and so on.

Figure 2.3b shows the waveforms in the time domain at the capacitor (V_{cap}) and at the output of the VFC (V_{out}), for $k = 4\ \mu S$, $V_{in} = 1$ V, $C_{int} = 5$ pF, $V_H = 0.8$ V, and $V_L = 0.4$ V, with a single supply of $V_{DD} = 1.8$ V. The output frequency is, according to (2.1), 1 MHz.

Note than when this VFC is supplied with symmetrical voltages, only the positive half of the voltage range can be converted into a frequency because this VFC cannot integrate negative input voltages. However, when it is single supplied, its input can swing from 0 to V_{DD}. Thus, in both cases, the theoretical input range is $[0, V_{DD}]$.

2.2.2 Asynchronous Charge-Balance VFC

Figure 2.4a shows the general scheme of a charge-balance VFC, and a typical implementation is shown in Fig. 2.4b. It consists of a voltage-to-current converter, a current integrator, a control circuit—usually made up of a comparator and a monostable (one-shot)—and a reference current source responsible for the charge balance.

As shown in Fig. 2.4c, the waveform at the output of the integrator is a two-part linear ramp: the first part lasts for time T_1, which depends on the input voltage, and the second one lasts a fixed time T_2, which corresponds to the pulse given by the monostable. The input voltage V_{in} is converted into a current I_{in}, then this current is integrated, charging the capacitor C_{int}. When the output of the integrator reaches the threshold voltage V_u of a single comparator, the comparator changes its state, triggering the monostable. At that moment, the precision monostable provides a pulse during a fixed time T_2. During this time, a reference current I_{ref}, $\left| I_{ref} \right| > \left| I_{in} \right|$, is subtracted from the circuit, subtracting a fixed charge from the capacitor, whereas the input current is continuously flowing during the discharge, so no input charge is lost. Once the pulse from the monostable ends, the cycle starts again, resulting in an output pulse rate that is accurately proportional to the rate at which the integrator charges from the input.

The iteration of this cycle gives a sawtooth wave at the integrator output, whose frequency is f_0. This output frequency is proportional to the sum of the charge time T_2 and the discharge time T_1, according to (2.2). That is, the output frequency depends on T_2, which is the pulse width given by the precision monostable; on I_{in}, the applied current proportional to the input voltage V_{in}; and on I_{ref}, the reference current, which is responsible for the charge balance.

$$f_0 = \frac{1}{T_1 + T_2} = \frac{I_{in}}{I_{ref}} \frac{1}{T_2} \tag{2.2}$$

This kind of VFC is more complex and demands more power than the multivibrator one and is able to achieve around 16 – 18 bits of linearity. Critical parameters are the pulse width of the monostable and the value of the reference current, which must be very stable [PAL01a]. At low frequencies, the stability of the voltage sources (V_{ref} and V_u) and the stability of the monostable (which mainly

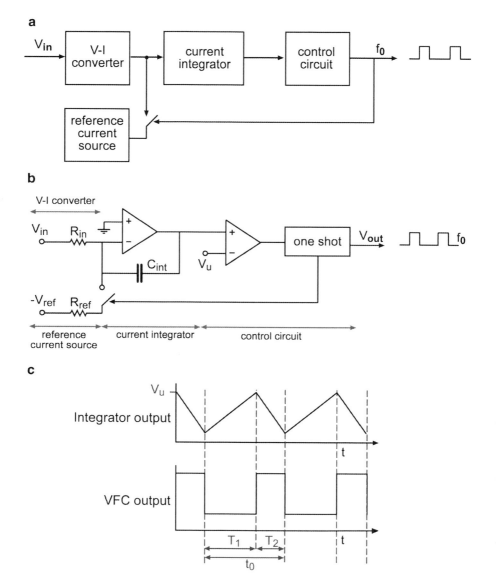

Fig. 2.4 Asynchronous charge-balance VFC: (**a**) block diagram, (**b**) typical implementation and (**c**) main voltage signals

depends on the capacitor stability) compromise the proper operation of the VFC. Temperature stability of the capacitor does not strongly affect the accuracy, but it does affect its dielectric absorption and leakages. At high frequencies second-order effects, such as switching transients in the integrator and in the monostable when being triggered shortly after the end of a pulse, affect the accuracy and the linearity [KES05].

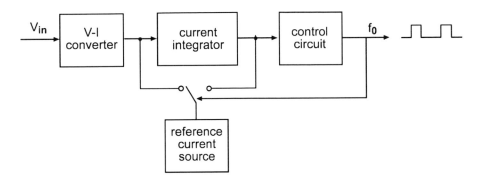

Fig. 2.5 Improved proposal of the asynchronous charge-balance VFC

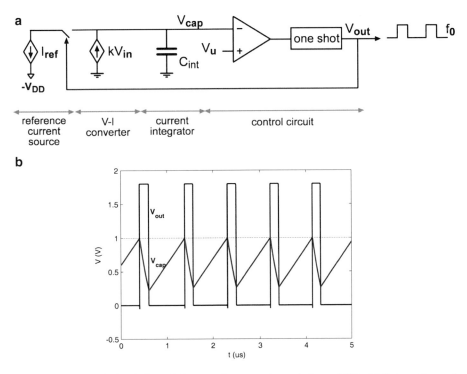

Fig. 2.6 Functional simulation of the asynchronous charge-balance VFC: (a) Schematics and (b) capacitor and output waveforms

To address the integrator transient problem, more recent charge-balance techniques introduce a changeover switch in the reference current source [ZUM08]: instead of having on/off transients, this current source is connected alternately to both sides of the integrator (see Fig. 2.5). With this change, the output stage of the integrator sees a constant load; most of the time the current from the source flows directly in the output stage. During the charge balance, it still flows in the output stage but through the integration capacitor.

The scheme shown in Fig. 2.6a was implemented to model the asynchronous charge-balance VFC in Fig. 2.4a. The input voltage V_{in} is converted into a current $I_{in} = kV_{in}$ with an ideal voltage-controlled current source. This current is integrated in a grounded capacitor C_{int}. The voltage across this capacitor is compared with a threshold voltage V_u in an ideal comparator. When the voltage across capacitor C_{int} reaches V_u, the output of the comparator changes and shoots the monostable. During T_2, the time that the monostable is active, a reference current I_{ref} is connected to the circuit, subtracting a fixed charge from the capacitor. When T_2 ends, the reference current is disconnected, starting the charging phase again. The output of the VFC is the output of the monostable.

Figure 2.6b shows the waveforms in the time domain at the integrating capacitor (V_{cap}) and at the output of the VFC (V_{out}), for $k = 10$ µS, $V_{in} = 1$ V, $C_{int} = 20$ pF, $I_{ref} = 50$ µA, $V_u = 1$ V, and $T_2 = 200$ ns with a symmetrical supply voltage of $V_{DD} = \pm 1.8$ V. The output frequency is 1 MHz according to (2.2).

Note that the required condition $|I_{ref}| > |I_{in}|$ to ensure the charge balance makes this type of VFCs to be typically biased with dual $\pm V_{DD}$ supply voltage, so that the input voltage swings from 0 to V_{DD}, whereas the reference current can be adequately generated using as a negative reference voltage $-V_{ref}$ and making the conversion into a current by means of a resistor R_{ref}, ($I_{ref} = -V_{ref}/R_{ref}$), as shown in Fig. 2.4b, which depicts the most widely used electronic implementation based on V-I converter with floating capacitor. For a single supply implementation, the input range would be reduced to [ΔV, V_{DD}], ΔV being the current source biasing voltage headroom.

2.2.3 Synchronous Charge-Balance VFC

The stability and the transient response of the monostable imply linearity problems in the charge-balance VFC. An alternative to the charge-balance VFC that solves this issue is the synchronous charge-balance VFC (SVFC), whose general diagram is shown in Fig. 2.7a: instead of a monostable, a bistable driven by an external clock is used. Thus, the linearity is increased, achieving around 18 bits and a high temperature stability, which makes them especially demanded for applications that require a high-resolution VFC [KES09]. A typical implementation of this kind of VFC is shown in Fig. 2.7b.

In this type of converters, as shown in Fig. 2.7c, the discharging phase does not start when the integrator output reaches the threshold voltage but with the next clock cycle. The output frequency expression is given by:

$$f_0 = \frac{I_{in}}{I_{ref}} f_{clk} \qquad (2.3)$$

Hence, the SVFC output is synchronized with the clock, so it is easier to process it with external or internal microcontroller counters. However, as the output pulses are aligned with clock pulses, the output pulses are not equally spaced, which

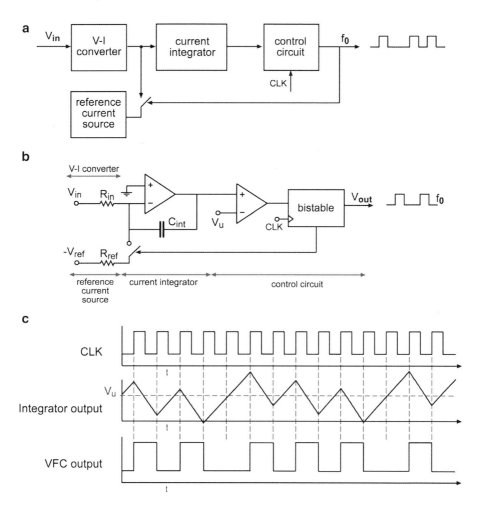

Fig. 2.7 Synchronous charge-balance VFC: (**a**) block diagram, (**b**) typical implementation and (**c**) main voltage signals

means that the output of a SVFC is not a pure tone like a conventional VFC, but contains components harmonically related to the clock frequency. This is a common cause of confusion when displaying the SVFC output on an oscilloscope because, in an asynchronous VFC, an increase in the input voltage is translated into an increase in the output frequency, whereas a change in a SVFC produces a change in the probability density of output pulses N and $N + 1$ clock cycles after the previous output pulse, which often is misinterpreted as severe jitter and a sign of a faulty device [KES05]. This might not affect the use of the SVFC as a part of an ADC, but it does when used as a precision oscillator. Another disadvantage appears close to subharmonic clock frequencies. This is due to the capacitive coupling of the clock into the comparator, which causes injection-lock effects, causing a small dead zone in its response [KES09].

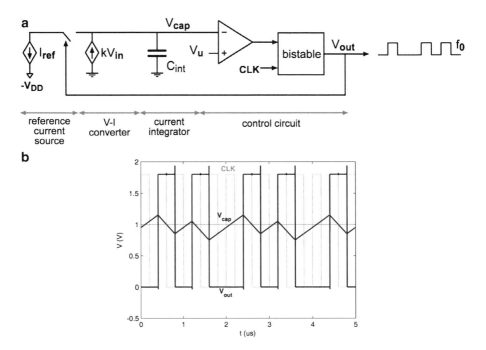

Fig. 2.8 Functional simulation of the synchronous charge-balance VFC: (**a**) Schematics and (**b**) capacitor, output and clock waveforms

Despite these disadvantages, the SVFC has better performances compared with the asynchronous approach, which makes it suitable for high resolution VFC applications and in multichannel systems because it removes problems of interference that can appear when having multiple asynchronous frequency signals [ZUM08].

The scheme shown in Fig. 2.8a was implemented to model the synchronous charge-balance VFC in Fig. 2.7a. As in the asynchronous VFC, the input voltage V_{in} is converted into a current $I_{in} = kV_{in}$ with an ideal voltage-controlled current source. This current is integrated in a grounded capacitor C_{int}. The voltage across the capacitor is compared with a threshold voltage V_u in an ideal single comparator. When the voltage across capacitor reaches V_u, the output of the comparator changes but, in contrast to what happens in the asynchronous approach, this time the current reference I_{ref} is not connected to the circuit until the next clock pulse comes, and therefore, the charge phase continues until the next clock pulse. Once the reference current is connected, a fixed charge is subtracted from the capacitor during a period of clock.

The waveforms at the capacitor and the output of the VFC are shown in the time domain in Fig. 2.8b, for $k = 10\ \mu S$, $V_{in} = 1\ V$, $C_{int} = 20\ pF$, $I_{ref} = 25\ \mu A$, $V_u = 1\ V$ and $f_{clk} = 2.5\ MHz$ with a symmetrical supply voltage of $V_{DD} = \pm 1.8\ V$. The output frequency is 1 MHz according to (2.3). Note that, as in the asynchronous charge-balance VFC, to ensure the condition $|I_{ref}| > |I_{in}|$, the VFC is usually biased with symmetrical $\pm V_{DD}$ supply voltage being the input voltage $[0, V_{DD}]$ and the reference current $I_{ref} = -V_{ref}/R_{ref}$. So, again the input range would be reduced to $[\Delta V, V_{DD}]$ for a single supply implementation, ΔV being the current source biasing voltage headroom.

2.3 Frequency-to-Code Conversion Methods

When used as a building block in an ADC system, a VFC exhibits excellent accuracy and linearity. Besides, although VFCs are not fast converters (they are slower than successive approximation devices [FRA10] but comparable to integrating ADCs), by using efficient frequency-to-code conversion methods, a good speed-accuracy trade-off can be obtained [BUR94].

There are many methods of frequency-to-code conversion, such as the standard counting method, the indirect method, the interpolation method, the method of recirculation, or the reciprocal counting method [KIR02b, YUR08]. However, when set against others, the simplicity, high performance, and universality of the standard direct counting method and the indirect counting method have contributed to their popularity. This is why they have become the preferred methods despite some restrictions and faults [KIR02b]. Therefore, these two methods are going to be studied in this section.

2.3.1 Standard Direct Counting Method (DCM)

This method is a frequency measurement technique, and it consists in counting a number of pulses N_x of unknown frequency f_x (with a period T_x) during a fixed gate time T_W.

$$N_x = \frac{T_W}{T_x} = T_W f_x \qquad (2.4)$$

Therefore, the unknown frequency f_x is given by:

$$f_x = \frac{N_x}{T_W} \qquad (2.5)$$

Figure 2.9 shows the DCM time diagram. When applying this method there are two main errors, the quantization error δN_x, and the reference frequency error δf_W.

The quantization error is due to the lack of synchronization of the beginning and the end of the gate time (T_W) with pulses f_x. The practical time where pulse counting is performed is T'_W, determined by values Δt_1 and Δt_2:

$$T'_W = N_x T_x = T_W + \Delta t_1 - \Delta t_2 \qquad (2.6)$$

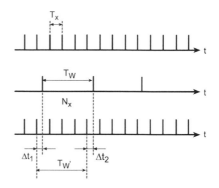

So that T_W can be rewritten as

$$T_W = N_x T_x - \Delta t_1 + \Delta t_2 = N_x T_x \pm \Delta t \tag{2.7}$$

Time intervals Δt_1 and Δt_2 can change independently of each other, varying Δt from 0 to T_x. Then, the maximum relative quantization error caused by the absence of synchronization is $\delta N_x = \pm 1$ [KIR02b].

Concerning the reference frequency f_W, two different errors should be considered: (1) $\delta f_{W,ref}$ is a systematic error caused by inaccuracy of the initial tuning and the long-term instability of the generator frequency and is a fixed error, and (2) $\delta f_{W,T}$ is the deviation of the real frequency from the nominal value because of temperature variations in a non-temperature-compensated crystal oscillator. Therefore, the error δf_W can be expressed as

$$\delta f_W = \delta f_{W,ref} + \delta f_{W,T} \tag{2.8}$$

Now, since (2.5) is an indirect measure, the frequency error can be calculated by applying the error propagation method, which results in the following expression:

$$\delta f_x = \left[f_W^2 (\delta N_x)^2 + N_x^2 (\delta f_W)^2 \right]^{1/2} \tag{2.9}$$

If N_x is written in terms of f_W, and considering that δf_W can be computed as $\delta f_W = f_W \, \Delta f_W$, where Δf_W is the f_W error per unit, the frequency error (2.9) related to f_x is given by

$$\frac{\delta f_x}{f_x} = \left[\frac{f_W^2}{f_x^2} (\delta N_x)^2 + (\Delta f_W)^2 \right]^{1/2} \tag{2.10}$$

Analyzing (2.10), the reference error term is fixed and it is independent on the reference frequency, whereas the term related to the quantization error increases as

Fig. 2.10 Indirect counting method time diagram [KIR02b]

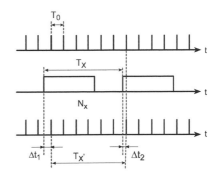

the unknown frequency decreases. In addition, focusing on δN_x, it can be negligible for high frequencies and significant for very low frequencies. Thus, this method is a right option to measure high frequencies. Effective methods for reducing the quantization error are: (a) opening a longer gate, kT_W, and (b) using weight functions; however, both choices result in an increase of the conversion time. One of the disadvantages of this classical method is the redundant conversion time in all frequency ranges, except for the nominal frequency [KIR02b].

One last thing that has to be taken into account is that, if the counter that is going to sum the number of pulses has a number of bits that is denoted by N, the gate time T_W required to have the maximum resolution depends on both the maximum and minimum frequencies to be measured, and is given by (2.11) [KLO12, PAL01a].

$$T_W = \frac{2^N}{f_{x,max} - f_{x,min}} \qquad (2.11)$$

2.3.2 Indirect Counting Method (ICM)

This method is a period measurement technique, and it is commonly used for low or extra low frequencies. In this case, a number of pulses N_x of a reference frequency f_W (with period T_W) are counted during m periods of the unknown frequency f_x (with period T_x), as given in (2.12).

$$N_X = m\frac{T_x}{T_W} \qquad (2.12)$$

Therefore, the unknown frequency f_x is given by

$$f_x = \frac{m}{N_x}f_W \qquad (2.13)$$

Figure 2.10 shows the ICM time diagram. There are two main errors associated, the quantization error δN_x and the reference frequency error δf_W.

The quantization error is due to the lack of synchronization of the beginning and the end of the gate time, T_x, with pulses f_W. The practical time where pulse counting is done is T_x' whose absolute value is determined by values Δt_1 and Δt_2.

$$mT_x' = N_x T_W = mT_X + \Delta t_1 - \Delta t_2 \tag{2.14}$$

Hence,

$$mT_X = N_x T_W - \Delta t_1 + \Delta t_2 = N_x T_W \pm \Delta t \tag{2.15}$$

Time intervals Δt_1 and Δt_2 can change independently of each other, varying Δt from 0 to T_W. Then, as in the DCM, the maximum relative quantization error caused by the absence of synchronization is $\delta N_x = \pm 1$ [KIR02b].

As has been explained in the previous method, for the reference frequency f_W two different errors have to be considered: (1) $\delta f_{W,ref}$ and (2) $\delta f_{W,T}$, so that the known frequency error δf_W is given by

$$\delta f_W = \delta f_{W,ref} + \delta f_{W,T} \tag{2.16}$$

and the final frequency error results in the following expression:

$$\delta f_x = \left[\left(\frac{f_x}{N_x} \right)^2 (\delta N_x)^2 + \left(\frac{m}{N_x} \right)^2 (\delta f_W)^2 \right]^{1/2} \tag{2.17}$$

If N_x is written in terms of f_W, and considering that δf_W can be computed as $\delta f_W = f_W \Delta.f_W$, where $\Delta.f_W$ is f_W error per unit, the frequency error (2.17), related to the working frequency, is given by

$$\frac{\delta f_x}{f_x} = \left[\left(\frac{f_x}{mf_W} \right)^2 (\delta N_x)^2 + (\Delta.f_W)^2 \right]^{1/2} \tag{2.18}$$

Analyzing (2.18), the reference error term is independent of the unknown error, and it is easy to see that the error grows as the unknown frequency f_x grows, and the error can be negligible for low frequencies. Thus, this method is the right option to measure low frequencies. Short time intervals cause a large quantization error. This can be reduced by: (a) increasing the reference frequency f_W and converting a greater number of intervals n or (b) with the interpolation method, which instead of an integer number of reference frequency periods filling out the converted time interval, fractional parts of this period reference pulse also are taken into account. The second method has a non-redundant conversion time, but it has a high quantization error in the medium and high frequency range [KIR02b].

In this case, if the counter that is going to sum the number of pulses has a number of bits denoted by N, the number of periods m of the unknown frequency f_x that the counter has to be active depends on the maximum frequency that is going to be measured, $f_{x,max}$, and is given by (2.19) [PAL01a].

$$m = \frac{f_{x,max}}{f_W} 2^N \tag{2.19}$$

2.4 Conclusions

In this chapter, the basic fundamentals of voltage-to-frequency converters have been introduced. First, to properly evaluate the performances of the different proposed VFCs, definition of key parameters has been performed. Next, the most common types of VFC, the multivibrator and the charge-balance VFC, have been reviewed, and finally, standard code-to-frequency conversion methods have been briefly explained.

This work is focused on the study of VFCs acting as a part of an analog-to-digital converter in sensor signal interface circuits within embedded microcontroller systems. In these systems, a quasi-digital converter can be used to digitize the sensor output signal, taking advantage of the microcontroller, which makes the final digitalization thanks to its internal clocks.

Particularly, in this book the VFCs are thought to be used in wireless sensor network nodes for environmental applications. For these networks, low-cost sensors that have usually around 8 – 10 bits accuracy are used. Therefore, the designed VFCs need only to have moderate accuracy. In addition, these networks are supplied with standard single supply batteries, typically 3 V. Thus, single-supply operation must be regarded, whereas low power is a must to extend battery life and, therefore, to maximize the operating life of the sensor node.

From the review of the most common VFC types in Sect. 2.2, and taking into account the requirements of moderate accuracy and single power supply, the proper starting point for the fully integrated VFC design seems to be a multivibrator VFC: it has a moderate 14-bit accuracy, which is good enough for our application, it is simpler than the charge-balance VFC, which means better low-voltage low-power compatibility, and its input range is not reduced with single voltage supply.

Reviewing the presented frequency-to-code conversion methods, the standard direct counting method is worse for low frequencies, whereas the indirect counting method does not work well for high frequencies. To use the proposed VFCs with an only frequency-to-code conversion method to simplify the microcontroller internal code, and considering that the VFC output frequencies have to be compatible with the microcontroller clock frequency, typically around 4 MHz, the standard direct counting method is the most suitable method. Therefore, to minimize the error in measuring low frequencies related to this method, an offset frequency will be added to the VFC to achieve an optimum output range $[f_{0,min}, f_{0,max}]$ for performing the digitalization.

References

[ALL12]   Definition of terms. http://www.allsensors.com/engineering-resources (2012)
[ANA12]   3 V/5 V low power, synchronous voltage-to-frequency converter. http://www.analog.com/static/importedfiles/data_sheets/AD7740.pdf (2012)
[BRY97]   Bryant, J.: VFC converters. Analog Dialog 23(2). http://www.analog.com/library/analogDialogue/Anniversary/3.html (1989)
[BUR94]   Voltage-to-frequency converters offer useful options in A/D conversion. Burr-Brown Application Bulletin, no. 91, USA. http://www.ti.com/lit/an/sbva009/sbva009.pdf (1994)
[FRA02]   Franco, S.: Signal generators. In: Design with Operational Amplifiers and Analog Integrated Circuits, 3rd edn. Tata McGraw Hill, New York (2002)
[FRA10]   Fraden, J.: Sensor characteristics. In: Handbook of Modern Sensors: Physics, Designs and Applications, 4th edn. Springer, New York (2010)
[HUI08]   Huijsing, J.H.: Interface electronics and measurement techniques for smart sensor systems. In: Meijer, G.C.M. (ed.) Smart Sensor Systems. Wiley, Chichester, UK (2008)
[KEN05]   Kenny, T.: Basic sensor technology. In: Wilson, J.S. (ed.) Sensor Technology Handbook. Elsevier, USA (2005)
[KES05]   Kester, W., Bryant, J.: Data converter architectures. In: Data Conversion Handbook. Newnes, USA (2005)
[KES09]   Kester, W., Bryant, J., MT-028 Tutorial: Voltage to frequency converters. Analog Devices Tutorials. http://www.analog.com/static/imported-files/tutorials/MT-028.pdf (2009)
[KIR02a] Kirinaki, N., Yurish, S., Shpak, N., Deynega, V.: Converters for different variables to frequency-time parameters of the electric signal. In: Data Acquisition and Signal Processing for Smart Sensors. Wiley, England (2002)
[KIR02b] Kirinaki, N., Yurish, S., Shpak, N., Deynega, V.: Methods of frequency-to-code conversion. In: Data Acquisition and Signal Processing for Smart Sensors. Wiley, England (2002)
[KLO12]   Klonowski, P., Application Note AN-276: Analog-to-digital conversion using voltage-to-frequency converters. Analog Application Notes. http://www.analog.com/static/imported-files/application_notes/185321581AN-276.pdf (2012)
[PAL01a] Pallás-Areny, R., Webster, J.G.: Digital and intelligent sensors. In: Sensors and Signal Conditioning. Wiley, New York (2001)
[PAL01b] Pallás-Areny, R., Webster, J.G.: Introduction to sensor-based measurement systems. In: Sensors and Signal Conditioning. Wiley, New York (2001)
[PEA80]   Pease, R.A., Application Note: V/F converter ICs handle frequency-to-voltage needs. National Semiconductor Application Notes. http://www.ti.com/lit/an-/snoa734a/snoa734a.pdf (1980)
[REV03]   Reverter, F., Jordana, J., Pallàs-Areny, R.: Program-dependent uncertainty in period-to-code converters based on counters embedded in microcontrollers. In: Proceedings of the 20th IEEE Instrumentation and Measurement Technology Conference (IMTC'03), vol. 2, pp. 977–980 (2003)
[SIL00]   de Silva, D.W.: Signal conditioning and modification. In: Vibration: Fundamentals and Practice. CRC Press, Boca Raton (2000)
[STO05]   Stork, M.: Sigma-delta voltage to frequency converter with phase modulation possibility. Turk. J. Electr. Eng. Comput. Sci. 13(1), 61–68 (2005)
[VID05]   Vidal-Verdú, F., Navas-González, R., Rodriguez-Vázquez, A.: Voltage-to-frequency converters. In: Chang, K. (ed.) Encyclopedia of RF and Microwave Engineering. Wiley, New Jersey (2005)
[YUR04]   Yurish, S.Y.: Sensors and transducers: frequency output vs voltage output. Sens. Transducers Mag. 49(11), 302–305 (2004)
[YUR08]   Yurish, S.Y.: Data acquisition for frequency- and time-domain sensors. In: Meijer, G.C.M. (ed.) Smart Sensor Systems. Wiley, Chichester, UK (2008)
[ZUM08]   Zumbahlen, H.: Converters. In: Linear Circuit Design Handbook. Newnes, USA (2008)

Chapter 3
Basic VFC Cells

Voltage-to-frequency converter circuit architectures based on multivibrator implementations basically consist on an input voltage-to-current converter followed by a bidirectional current integrator driven by a control circuit, usually a voltage window comparator (VWC) or a Schmitt trigger (ST).

In this chapter an introduction to each of these three basic cells is done and some implementations are advanced pursuing the requirements set by the driving application of this work, which remember mainly are: (1) low voltage supply, to be powered with the batteries used in WSN nodes, (2) low power consumption, to maximize batteries life, (3) rail-to-rail input operation, to take advantage of the maximum achievable conversion resolution, (4) output frequency range compatible with WSN µC clocks, typically $f_{clk} = 1 - 4$ MHz, and (5) temperature and supply voltage-independence, to maintain constant VFC performance characteristics. In addition, to facilitate a system-on-chip solution, thus reducing the total area, increasing the operation velocity, avoiding wiring faults and parasitics and, especially, reducing the cost, the implementation will be done in a CMOS technology. In particular, the chosen technology is a standard CMOS process, provided by United Microelectronics Corporation (UMC), with minimum gate length of 0.18 µm, P-substrate/N-well, six metal layers and one polysilicon layer [UMC12].

Thus, voltage-to-current converters are studied in the first section; next, bidirectional current integrators are explained in the second section and control circuits are introduced in the third section. Bias circuits, also presented in this chapter, are considered in the fourth section. Finally, as a closing section, general conclusions are given.

3.1 V-I Converters

The voltage-to-current (V-I) converter is the input stage of the VFC. This cell, which is a basic building block in many analog and mixed signal designs, such as multipliers, continuous-time Gm-C filters, data converters, high-performance

C.A. Murillo et al., *Voltage-to-Frequency Converters: CMOS Design and Implementation*,   35
Analog Circuits and Signal Processing, DOI 10.1007/978-1-4614-6237-8_3,
© Springer Science+Business Media New York 2013

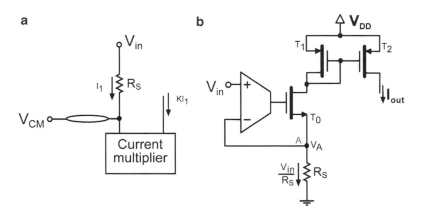

Fig. 3.1 Conventional V-I converters: (**a**) with passive resistors at the input and (**b**) conventional OTA-MOS-resistance V-I converter

sensor interfaces or variable gain amplifiers, is critical because the overall system performance depends largely on the V-I converter features. This leads to the need for a time-, temperature-, and voltage level-independent transconductance, with a high linear range and an appropriate bandwidth.

Especially critical is the input range: in a VFC, the V-I input range determines the system input operating range. Therefore, attaining rail-to-rail operation for the V-I converter is a must because, if this feature is not accomplished, there will be a loss in the maximum allowable VFC signal-to-noise ratio (SNR). As a consequence, we will not take advantage of the full sensitivity in the subsequent DCM frequency-to-code conversion and, for a given time conversion gate T_W, effective resolution is lost. However, with the reduction of power supply voltage motivated by the down-scaling of the CMOS processes, V-I converters, as well as other essential analog building blocks, lose a significant amount of operating range. Thus, they need to be redesigned in order to keep high performances to not limit the system operation.

Besides, the preferred alternative to have a highly linear transconductor is the use of passive resistors for performing the V-I conversion. The simplest choice uses passive resistors at the V-I input terminals tied to a virtual ground by means of a nullator, which is a two-terminal device that has no voltage drop between its terminals, and no current flows across it, as shown in Fig. 3.1a. In the recent literature, a number of CMOS rail-to-rail V-I converters are reported based on this strategy [HAS11, LOP07, SHU04, SRI05]. As an example, in [HAS11] a low-voltage (1.0 V) low-power (0.69 mW) rail-to-rail V-I converter presents, for an input resistor $R_{in} = 50$ kΩ, a transconductance of 20 μS, a bandwidth of 238 MHz, and a $THD = -52$ dB for 1 V_{pp} − 1 MHz signal. Nevertheless, these V-I converters always exhibit a rather small input impedance, as a compromise between area consumption and resistor values. Consequently, an auxiliary rail-to-rail input buffer must be added to properly process the signal, adding complexity to the system and increasing the total power consumption and noise.

Conversely, the conventional V-I converter based on an operational amplifier (OA) or a transconductance amplifier (OTA) driving a MOS (T_0) and a grounded linear resistor (R_S) in a negative feedback loop [WAN06a], as shown in Fig. 3.1b, is highly linear while offers a high impedance input node.

To analyze more thoroughly the behavior of this and the subsequent V-I converters, the OTA is modelled by a transconductance G and a dominant pole (formed by the output capacitance C and resistance R), while for the transistors simplified small-signal model have been used, where parameters denote their usual meaning using the subscript $_i$ for transistor T_i. All these detailed analysis, not shown here for easy the reading, are included in Appendix B.

With this simple modelling, for this structure the input voltage V_{in} is buffered to node A according to the relationship:

$$\left.\frac{V_A}{V_{in}}\right|_{\omega \to 0} \approx \frac{GRg_{m0}R_L}{1 + g_{m0}R_L + g_{m0}R_LGR} \approx \frac{GR}{1 + GR} \approx 1 \qquad (3.1)$$

where R_L denotes the parallel connection of R_S and $1/g_{mb0}$. This buffered voltage V_A is converted into a current I_{in} across resistor R_S, achieving a linear V_{in}-to-I_{in} conversion inversely proportional to R_S given by $I_{in} = V_A/R_S$.

The generated current I_{in} is driven to the output through a simple current mirror formed by transistors T_1–T_2, so that the transconductance of the system can be expressed as

$$G_M|_{\omega \to 0} \approx \frac{1}{R_S}\frac{(W/L)_2}{(W/L)_1}\left(1 + \frac{V_{DS2} - V_{DS1}}{V_E}\right) = \frac{1}{R_S}\frac{(W/L)_2}{(W/L)_1}(1 + \varepsilon) \qquad (3.2)$$

where V_E is the Early voltage and the rest of parameters have their usual meaning. As reflects (3.2), and it is well known, the main deviation ε of the current copy from its ideal value is due to unequal T_1 and T_2 drain-source voltages and channel length modulation, since other sources of error due to imperfect geometrical matching, technological parameter mismatch and parasitic resistances can be minimized by proper layout techniques. The input voltage for this current mirror is $V_{in,min} = |V_{th1}| + V_{DS,sat1}$, and the minimum output voltage needed to keep transistor T_2 working in saturation region is $V_{out,min} = V_{DS,sat2}$ [CHE05, COR03, GRA10].

The current copy can be improved in a simple way keeping compatibility with low-voltage biasing requirements by using a high-swing cascode mirror [CRA92, JOH97], shown in Fig. 3.2a. In this mirror, V_{DS1} and V_{DS2} are controlled by the gate of transistors T_{1C} and T_{2C}, respectively, so that the matching between these transistors ensures identical drain-source voltages, and the output resistance is increased up to $r_{out} = g_{m2C}r_{02}r_{02C}$. This results in a negligible deviation $\varepsilon \approx 0$. The input voltage for this high-swing cascode current mirror is given by $V_{in,min} = V_{DS,sat1} + |V_{th1}|$ and the minimum output voltage is $V_{out,min} = V_{DS,sat2} + V_{DS,sat2C}$ [GRA10, RAM04].

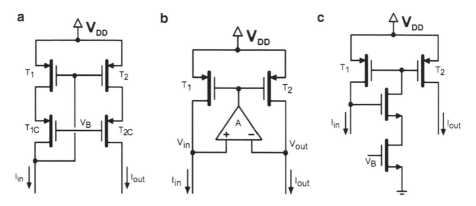

Fig. 3.2 (**a**) High-swing cascode mirror and low-voltage current mirrors: (**b**) drain voltage balanced and (**c**) level shifted

However, both the simple and the high-swing cascode current mirror require a significant minimum input voltage V_{in}, thus reducing noticeably the operating range of the derived V-I converter to $V_{DD} - V_{in,min}$, which makes the use of these mirrors unsuitable in low-voltage applications. An increase in the V-I operating range can be achieved using other low-voltage current mirror topologies. One widely used proposal is the error amplifier based current mirror, shown in Fig. 3.2b. The input current I_{in} is introduced in the circuit and mirrored by transistors T_1–T_2, whose gate voltage is driven by an error amplifier A which ensures that $V_{DS1} = V_{DS2}$ although V_{out} changes. In this case the minimum input voltage is $V_{DS,sat1}$ [RAM04, VIE08, YOU97], but this solution requires the additional error amplifier A and it must have high gain to guarantee proper circuit operation, therefore increasing the area and the power consumption. The proposal shown in Fig. 3.2c presents a low voltage current mirror that employs a level-shifter between the gate and the drain of the input transistor, in order to reduce the input voltage requirements, providing also a minimum input voltage of $V_{DS,sat1}$. However, this current mirror does not improve the current copy compared to the simple current mirror [DOU04, RAM94, VIE08]. Other current mirrors can be found in [BLA95, DOU04, RAM04, RAM94, VIE08]. In general, all these schemes have the disadvantage of requiring not trivial extra circuit, implying an extra area and power consumption penalty.

Therefore, this section explores new V-I converter designs to achieve full input range while maintaining the linearity of the conventional V-I structure.

3.1.1   Enhanced V-I Converters

To extend the VFC voltage operating range close to the supply voltage preserving linearity, a V-I conversion circuit based on an OTA/common-source amplifier configuration can be used [AZC11, CAL10, VIE07]. This scheme, shown in Fig. 3.3, is

Fig. 3.3 OTA-common-source V-I converter

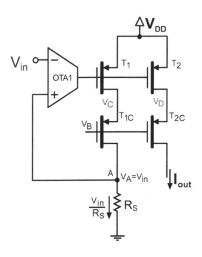

basically a two-stage operational amplifier, and it is a known topology in voltage regulator design since the output V_A can swing all the way up to V_{DD} [BAK98]. However, although the V_{in}-V_A buffering can cover all the range between the supply voltages, the input range of this V-I converter is still limited to $V_{DD} - V_{DS,sat1} - V_{DS,sat1C}$ because, for voltages above this value, transistors T_1 and T_{1C} enter triode region, losing the linear relationship between V_{in} and I_{out}. Thus, with respect to the conventional V-I converter of Fig. 3.1b, the input range is widened from $V_{in} < V_{DD} - V_{DS,sat1} - |V_{th1}|$ to $V_{in} < V_{DD} - V_{DS,sat1} - V_{DS,sat1C}$, that is, the input voltage range is the same as using an error amplifier based cascoded current mirror (Fig. 3.2b), but this one has the advantage that does not use additional circuits. In order to improve the current copy, cascode transistors T_{1C}–T_{2C} are introduced, chosen to be $T_{1C} = T_1$ and $T_{2C} = T_2$.

Note that, in this scheme, the feedback loop seems to be positive; however, since the amplifier T_1 is inverting, the negative feedback is accomplished by connecting the output to the positive input of the OTA.

The input voltage V_{in} is buffered to V_A according to

$$\left.\frac{V_A}{V_{in}}\right|_{\omega \to 0} = \frac{g_{m1}(r_{01}\|R_S)GR}{1 + GR(r_{01}\|R_S)g_{m1}} \approx 1 \tag{3.3}$$

This buffered voltage is converted into a current I_{in} across resistor R_S, achieving a linear V_{in}-to-I_{in} conversion inversely proportional to R_S given by $I_{in} = V_A/R_S$. The generated current I_{in} is sensed by transistors T_1–T_{1C} and directly conveyed to the output by transistors T_2–T_{2C}, so that the transconductance can be expressed as

$$\left.\frac{I_{out}}{V_{in}}\right|_{\omega \to 0} = g_{m2}\frac{GR}{1 + GR(r_{01}\|R_S)g_{m1}} \approx \frac{g_{m2}}{g_{m1}}\frac{1}{R_S} = \frac{(W/L)_2}{(W/L)_1}\frac{1}{R_S} \tag{3.4}$$

We next consider the system frequency response. Looking to the obtained poles derived from a straightforward analysis, it cannot be assured that a dominant pole exists. Therefore, in order to avoid undesirable peaks in the closed loop frequency response and underdamped oscillations, a conventional $R_C C_C$ compensation network is introduced between the OTA output and node A which allows to make a dominant pole approximation. Thus, the buffered voltage V_A expression is given by

$$\frac{V_A}{V_{in}}(s) \approx \frac{g_{m1} G R_S}{(g_{m1} G R_S + s C'_T)} \tag{3.5}$$

and the transconductance can be expressed as

$$G_M(s) \approx \frac{1}{R_S} \frac{(W/L)_2}{(W/L)_1} \frac{g_{m1} G R_S}{(g_{m1} G R_S + s C'_T)} \tag{3.6}$$

where $C'_T = C + C_{gs1} + C_{gs2} + C_C$.

These equations reveal that, in fact, there is a compromise in the dimensions of the output transistors T_1–T_{1C}. To have a maximum range in the V_{in}-V_A buffering, $V_{DS,1}$ is desired to be minimum, and therefore $(W/L)_1$ has to be set as large as possible. In addition, the larger the ratio $(W/L)_1$, the more valid the approximation $V_A/V_{in} \approx 1$ made in (3.3). However, the system bandwidth BW depends on the ratio $(W/L)_1$:

$$BW = \frac{g_{m1} G R_S}{C + C_C + C_{gs1} + C_{gs2}} = \frac{\sqrt{2 I_{out} (W/L)_1} G R_S}{C + C_C + 2 C_{OX} (W/L)_1 / 3 + C_{gs2}} \tag{3.7}$$

Thus if $(W/L)_1$ is made larger, bandwidth decreases.

Two V-I converters based in the scheme of Fig. 3.3 are next presented. The first one is a preliminary design in 0.18 μm with a single power supply of 1.8 V. The second one is an optimized design implemented in the same technology targeting low-voltage low-power operation, supplied with 1.2 V.

3.1.1.1 VIC 1: Enhanced V-I Converter

The scheme of the first proposed V-I converter is shown in Fig. 3.4. The OTA is made up of a simple differential pair M_1–M_4 fed by a current source $2I_B$, with all transistors working in the saturation region.

In this case, $G = g_{m2\_O}$, $R = r_{02\_O} \| r_{04\_O}$ and $C = C_{gd2\_O} + C_{gd4\_O}$, where the subscript $_O$ indicates that these parameters belong to the OTA. Thus, (3.6) becomes

$$G_M(s) \approx \frac{1}{R_S} \frac{(W/L)_2}{(W/L)_1} \frac{g_{m1} g_{m2\_O} R_S}{(g_{m1} g_{m2\_O} R_S + s C'_T)} \tag{3.8}$$

Fig. 3.4 Scheme of the
proposed VIC 1 converter

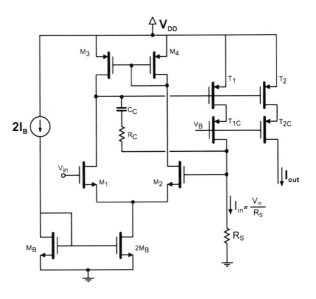

Table 3.1 Transistor sizes
of the VIC 1 converter

Transistor	W/L (μm/μm)
M_1–M_2	1/1.2
M_3–M_4	30/1.2
T_1–T_{1C}	50/1.2
T_2–T_{2C}	50/1.2

The input operating range of this converter is expected to be $V_{DS,sat2MB}$ + V_{th1} + $V_{DS,satM1}$ < V_{in} < V_{DD} − $V_{DS,satT1}$ − $V_{DS,satT1C}$. The lower limit is imposed by transistors $2M_B$ and pair transistor M_1 being in saturation. The upper limit is imposed by common-source transistors T_1–T_{1C}, working in the saturation region.

The VIC 1 scheme in Fig. 3.4 has been designed in the 0.18-μm CMOS technology from UMC with a single supply of V_{DD} = 1.8 V. Transistor sizes are given in Table 3.1. The scaling factor K between transistors T_1 and T_2 is set to K = 1. Transistors T_1–T_{1C}–T_2–T_{2C} have L = 1.2 μm to avoid short channel effects. To have a good matching between the differential pair transistors M_1–M_2, each of them are made as the parallel connection of two transistors with same length and half width using a common-centroid layout technique, including dummy transistors that ensure that the unity transistors of the matched transistors see the same adjacent structures.

The resistor that makes the V-I conversion is set to R_S = 20 kΩ and it is implemented by a high-resistivity polysilicon (HRP) layer. Theoretically, the variation on the nominal value of an integrated resistance can be up to 20 %. Hence, this fact can result in a significant error in the transfer function and has to be considered as a critical issue to analyze. Thus, R_S has been designed with twice the minimum width to minimize the deviation from the nominal value and a replica

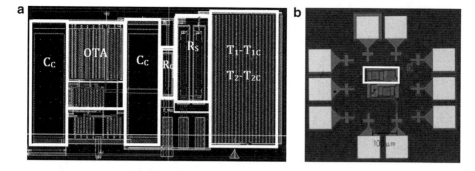

Fig. 3.5 (**a**) Layout and (**b**) microphotograph of the VIC 1 converter

Table 3.2 Main performances of the VIC 1 with a load resistance of $R_L = 10\,k\Omega$

Parameter	Value
Technology	0.18 μm CMOS
Supply voltage	1.8 V
Transconductance	50 μS
G_M deviation (T = 27 °C)	5.8 %
Input range	0.7 – 1.5 V
BW	38.2 MHz
THD (0.7 V_{pp}@1 kHz)	−37.1 dB
Input offset voltage	7.3 mV
Power consumption	315 μW

resistance cell has been included in the first 0.18-μm CMOS run to study its behavior. The fabricated resistors have been measured for ten different dies using a four wire technique, being measured with a Keithley 2010 multimeter and a 2000-Scan scanner card, obtaining a maximum deviation of 3.2 % with respect to its nominal value and a 0.09 % dispersion between the measured resistors. Therefore, no error due to the integrated resistance deviation is going to be considered. If a significant error would be achieved, at the end, this deviation would be translated into an error in the sensitivity of the VFC that has to be trimmed in the μC.

The compensation capacitor is implemented with a metal-insulator-metal (MIM) capacitor whose value is $C_C = 0.8\,pF$ and the compensation resistor is made by an HRP resistor of value $R_C = 5\,k\Omega$. The biasing current is set to $2I_B = 5\,\mu A$ and it is introduced into the circuit through a simple current mirror formed by transistors M_B with sizes 10 μm/1.2 μm.

The layout of the core and the die photograph are shown in Fig. 3.5, and its main performances are summarized in Table 3.2.

Figure 3.6 shows the normalized G_M at room temperature measured with a load resistor of $R_L = 10\,k\Omega$. The maximum G_M deviation is of 5.8 % over a (0.7, 1.5 V) range. Figure 3.7 shows the output current I_{out} vs. the input voltage at different temperatures, measured with a Fitoterm 22E thermal chamber from Aralab.

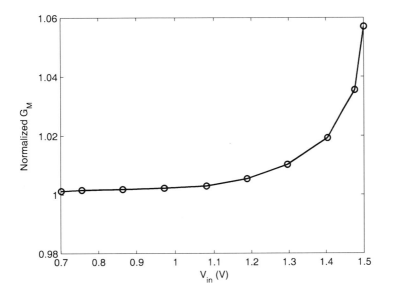

Fig. 3.6 Normalized G_M of the VIC 1 at room temperature

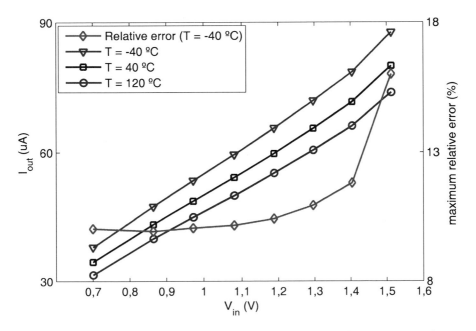

Fig. 3.7 VIC 1 output current vs. input voltage at different temperatures and the maximum relative error ($T = +120\ ^\circ C$)

The VIC 1 is highly linear (0.085 % linearity error), and exhibits an enhanced input range with respect to the conventional approach using the same OTA and cascode output stage. However, when set against temperature variations, the total transconductance does not remain constant, varying ± 10 % with respect to the room temperature transconductance, mainly due to the resistor R_S temperature dependence.

3.1.1.2 VIC 2: Subthreshold Enhanced V-I Converter

The second scheme is the same than in the previous approach, shown in Fig. 3.4, but in an attempt to optimize power consumption and input range, all transistors in the OTA work in the subthreshold region. The main performances of weak inversion region are the following [COM04, ENZ96, STE08, VIT09]: the ratio g_m/I_D is maximum, thus achieving higher gains for the same current; $V_{DS,sat}$ is smaller, allowing high dynamics and offset voltages are lower; on the other hand, the slew-rate is lower; current matching is poor; and frequency operation is lower. Biasing transistors in the subthreshold region is widely used as a power consumption reduction technique [CAR04, CHA10, COR06, ROD02].

As weak inversion bias is poor for current mirror matching compared to strong inversion, the current mirror will remain in saturation region.

In this converter, the expected input operating range is given by $V_{DS,sat2MB} + V_{th1} + nV_T\ln(I_B/I_{DO}) < V_{in} < V_{DD} - V_{DS,satT1} - V_{DS,satT1C}$, where now $V_{DS,sat2MB} = 4V_T$, n is the subthreshold slope factor, V_T the thermal voltage and I_{DO} the off-drain current, having all other parameters their usual meaning. The lower limit is imposed by subthreshold transistors $2M_B$ and M_1 being in saturation. The upper limit is imposed by the strong inversion common-source T_1–T_{1C} being in saturation region to provide a good current copy. Note that transistor $2M_B$ working in the triode region is not a serious drawback since the differential pair single output operation through M_3–M_4 current mirror cancels the common mode input signal contribution, so, the common mode rejection ratio (CMRR) is moderate but sufficient, thus being the input limit approximately $V_{th1} + nV_T\ln(I_B/I_{DO})$.

For this topology, $G = g_{m2\_O} = I_B/nV_T$, $R = r_{02\_O}||r_{04\_O}$ and $C = C_{gd2\_O} + C_{gd4\_O}$, where the subscript $_O$ again indicates that these parameters belong to the OTA. Thus, (3.6) can be rewritten as

$$G_M(s) \approx \frac{1}{R_S} \frac{(W/L)_2}{(W/L)_1} \frac{g_{m1}g_{m2\_O}R_S}{(g_{m1}g_{m2\_O}R_S + sC_T')} \tag{3.9}$$

Considering now the temperature drift, the dependence of the output current with temperature is mainly due to the resistor R_S that makes the V-I conversion. Therefore, if this resistor is implemented with a specific resistive layer among those provided by the integrating technology, for instance with a HRP layer as in the previous case, high temperature drifts are expected due to the resistance variation with temperature, which is given by

Table 3.3 Thermal coefficients of the R_{PND}, R_{HRP} and R_S resistors and their sheet resistances at room temperature

	$TC_1 (\cdot 10^{-3\circ}C^{-1})$	$TC_2 (\cdot 10^{-7\circ}C^{-2})$	$R_{\square} (\Omega/\square)$
R_{PND}	1.184	7.310	158
R_{HRP}	−0.834	13.0	1039
R_S	−0.0268	9.08	510.4

$$R(T) = R_0 \left(1 + TC_1(T - 25) + TC_2(T - 25)^2 \right) \qquad (3.10)$$

where R_0 is the resistor value at room temperature and TC_1 and TC_2 are the first and second order temperature coefficients, respectively, whose values appear in Table 3.3.

Hence, a temperature independent resistor has to be implemented. This is achieved by the serial connection of two resistors R_N and R_P with opposite T-coefficients, TC_N and TC_P, which provide the suitable first and second thermal coefficients TC_i, $i = 1, 2$ given by (3.11).

$$TC_i = TC_{iN} \frac{\beta}{\beta + 1} + TC_{iP} \frac{1}{\beta + 1} \qquad (3.11)$$

where $\beta = R_{0N}/R_{0P}$ is the ratio between resistances at room temperature [GRE07]. The resistor with negative T-coefficient is implemented with a HRP layer ($R_N = R_{HRP}$), the resistor with positive T-coefficient is implemented by means of a P^+ non-salicide diffusion (PND) layer ($R_P = R_{PND}$) and the appropriate ratio R_N/R_P that reduces the temperature dependence is $\beta = 1.5$. With this selection, the first and second order temperature coefficients of the composite resistor R_S are given in Table 3.3, as well as those corresponding to the R_{HRP} and R_{PND} resistors. To check their behavior, a set of ten replica composite resistors of 20 kΩ were fabricated as a serial combination of $R_{HRP} = 12$ kΩ and $R_{PND} = 8$ kΩ. These resistors were measured by using the 4-wires technique obtaining a maximum deviation of 4.1 % with respect to its nominal value and a 0.13 % dispersion between the measured resistors. Therefore, as in VIC 1, we will consider this component error-free. The composite resistors were next tested against temperature variations. Figure 3.8 shows the variation of a resistor with nominal value of 20 kΩ, made up with a HRP resistor, that is, uncompensated resistor, and of a 20 kΩ compensated composite resistor ($R_{HRP} + R_{PND}$) over a temperature range of (-40, $+120$ °C). The resistance varies less than 1.6 % over all the temperature range, which proves that the selected temperature compensation technique is correct.

Transistor sizes of this V-I converter are given in Table 3.4. For its implementation, same layout techniques than in VIC 1 converter have been followed. The resistor that realizes the V-I conversion is also set to $R_S = 20$ kΩ ($R_{HRP} = 12$ kΩ and $R_{PND} = 8$ kΩ). The biasing current is $2I_B = 0.5$ μA, ten times smaller than in VIC 1, and it is introduced into the circuit through a simple current mirror NMOS M_B (10 μm/1.2 μm) with 1:2 gain. The compensation MIM capacitor is set to $C_C = 0.4$ pF and, to reduce the integrating area, the compensation resistor R_C is the

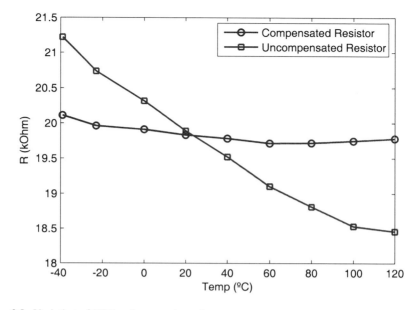

Fig. 3.8 Variation of HRP and composite resistances versus temperature

Table 3.4 Transistor sizes for the VIC 2 converter

Transistor	W/L (μm/μm)
M_1–M_2	60/1.2
M_3–M_4	30/1.2
T_1–T_{1C}	50/1.2
T_2–T_{2C}	50/1.2

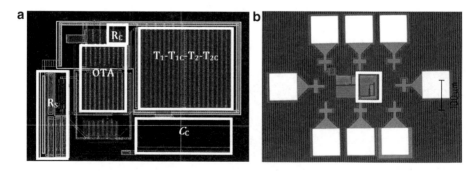

Fig. 3.9 (**a**) Layout and (**b**) microphotograph of the VIC 2 converter

parallel connection of a NMOS (2 μm/1.4 μm) and a PMOS (4 μm/1.4 μm) transistors with gate voltages $V_{GN} = V_{DD}$ and $V_{GP} = 0$ V, so that the total equivalent resistance R_C is kept approximately constant as V_{in} swings from 0 to V_{DD} [MAL01].

The layout and the die microphotograph of this converter are shown in Fig. 3.9, and its main performances are summarized in Table 3.5.

Table 3.5 Main performances of the subthreshold enhanced V-I converter with a load resistance of $R_L = 10 \text{ k}\Omega$

Parameter	Value
Technology	0.18 μm CMOS
Supply voltage	1.8 V
Transconductance	50 μS
G_M deviation	4.6 %
G_M deviation ($-40, +120$ °C)	6.2 %
Input range	0.1 – 1.6 V
BW	5.35 MHz
THD (1 V_{pp}@1 kHz)	-36.8 dB
Input offset voltage	13.8 mV
Power consumption	300 μW

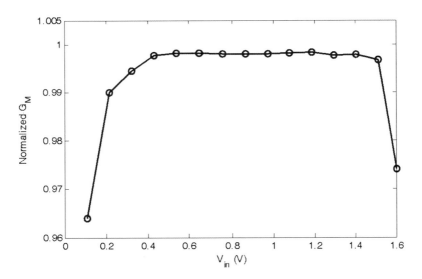

Fig. 3.10 Normalized G_M of the VIC 2 at room temperature

Figure 3.10 shows the normalized transconductance G_M at room temperature measured with a load resistor of $R_L = 10 \text{ k}\Omega$. The maximum G_M deviation is 4.6 %, achieving high linearity in the V-I conversion (0.037 % linearity error), over an input range of (0.1, 1.6 V). Figure 3.11 shows the output current I_{out} vs. the input voltage V_{in} for different temperatures, together with the maximum relative error corresponding to $T = -40$ °C. Thanks to the introduction of the composite resistance, the transconductance of the system varies ± 1.3 % over a $(-40, +120$ °C), with respect to the room temperature transconductance.

Note that bandwidth of the VIC 2 is smaller than that of VIC 1, since the OTA is working on subthreshold region. However, the power consumption has not been considerably reduced. In fact, just a slight reduction has been achieved. This is

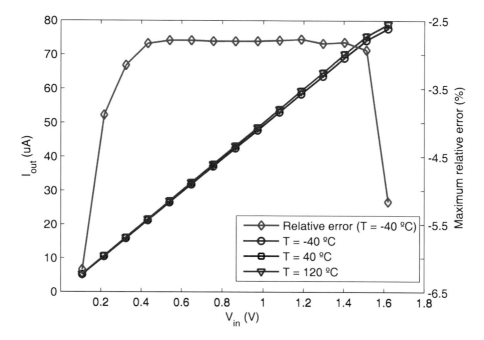

Fig. 3.11 VIC 2: output current vs. input voltage at different temperatures and the maximum relative error ($T = -40\,°C$)

because most of the power consumption is due to the R_S branch used for the V-I conversion. Thus, for the maximum input voltage and the same resistor R_S, the power consumption of both VIC 1 and VIC 2 are comparable. The lower limit of the input range is extended from 0.7 to 0.1 V, while the upper is maintained equal to that of VIC 1. This lower limit can be down to 0 V using a rail-to-rail OTA, but the upper range is limited to $V_{DD} - V_{DS,satT1} - V_{DS,satT1C}$, limitation that can be still critical in deep submicrometer CMOS technologies operating from sub-volt supplies. Thus, further work towards rail-to-rail operation is due.

3.1.2   Rail-to-Rail V-I Converters

The previously presented structures have significantly extended the input range when set against the conventional approach of Fig. 3.1b, preserving a highly linear relationship between the input voltage and the output current. In addition, VIC 2 explores subthreshold region operation, widening the operating range and slightly reducing the power consumption. Therefore, based on the VIC 2, the proposed solution to achieve a full rail-to-rail input range is presented in the general scheme of Fig. 3.12.

The idea is to reduce the voltage V_A across the conversion resistor R_S from V_{in} to αV_{in}, being $0 < \alpha < 0.9$, in order to keep transistor T_1 working in the saturation region over the complete input voltage range. To accomplish this voltage

Fig. 3.12 General scheme of
the rail-to-rail V-I converters

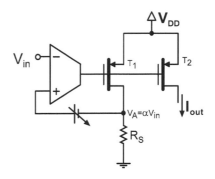

attenuation, a floating dynamic battery is introduced between the OTA non-inverting input and the resistor R_S. In this way, assuming a rail-to-rail input, the output current mirroring will cause no reduction in the V-I operating range. So, three new approaches that extend the operating range to rail-to-rail are next introduced and analyzed. These are based on feedforward voltage attenuation (FFVA), feedback voltage attenuation (FBVA) and feedforward current attenuation (FFCA). All have been designed in the 0.18-μm CMOS technology from UMC with a single supply of $V_{DD} = 1.2$ V and use the OTA-common source amplifier, thus being valid the analysis obtained in the previous section. However, instead of a simple differential pair, a rail-to-rail OTA with two complementary input stages working in subthreshold region is employed.

The scheme of the OTA, formed by transistors M_1–M_{12} fed by a current $2I_B$, and its transistor sizes are shown in Fig. 3.13. It can be described by parameters given in Table 3.6 for each of the three OTA operation regions: when NMOS differential pair is active, when PMOS differential pair is active, and when both of them work.

3.1.2.1 FFVA: Feedforward Voltage Attenuation V-I Converter

The feedforward voltage attenuation (FFVA) V-I converter, shown in Fig. 3.14 is the simplest proposal. In this approach, to attenuate the input voltage, a rail-to-rail input voltage divider is used before the main V-I converter. This input attenuator is made up of an auxiliary OTA (OTA$_{aux}$) loaded with a resistive voltage divider formed by linear resistors R_1 and R_2.

Thus, the FFVA converter basically consists of two cascaded enhanced V-I converters. Therefore, equations obtained in the previous section can be applied directly. First, the input voltage V_{in} is buffered to V_{out1} by the OTA$_{aux}$–T_{A1}–T_{AC1} voltage follower with rail-to-rail input–output operation. Cascode configuration is used to have a modular circuit, because cascode transistors will be used in the second stage to improve the current copy.

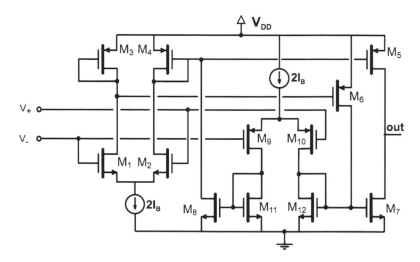

Fig. 3.13 Schematic view of the rail-to-rail OTA used in the proposed rail-to-rail V-I converters and its transistor sizes

Table 3.6 Rail-to-rail OTA parameters depending on the operation region

Region	G	R	C
$V_{in} < V_T ln\left(\frac{I_B}{I_{DO}}\right) + V_{th} + 4V_T$	$g_{m9} = \frac{I_B}{nV_T}$	$\frac{1}{I_B(\lambda_5+\lambda_7)}$	$C_{gd5} + C_{gd7}$
$4V_T < V_{in} < V_{DD} - 4V_T$	$g_{m1} + g_{m9} = \frac{2I_B}{nV_T}$	$\frac{1}{2I_B(\lambda_5+\lambda_7)}$	$C_{gd5} + C_{gd7}$
$V_{in} > V_{DD} - nV_T ln\left(\frac{I_B}{I_{DO}}\right) - V_{th} - 4V_T$	$g_{m1} = \frac{I_B}{nV_T}$	$\frac{1}{I_B(\lambda_5+\lambda_7)}$	$C_{gd5} + C_{gd7}$

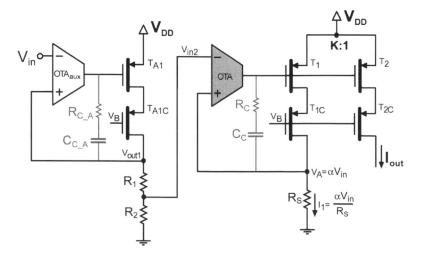

Fig. 3.14 Feedforward voltage attenuation V-I converter (FFVA)

The buffered voltage, according to (3.3), is given by

$$\left.\frac{V_{out1}}{V_{in}}\right|_{\omega \to 0} = \frac{g_{m1\_A}\left(r_{01\_A}||(R_1 + R_2)\right)G_{\_A}R_{\_A}}{1 + G_{\_A}R_{\_A}g_{m1\_A}\left(r_{01\_A}||(R_1 + R_2)\right)} \approx 1 \qquad (3.12)$$

where the subscript $_{\_A}$ denotes the parameters of the auxiliary elements OTA_{aux}-T_{A1}-T_{AC1} given in Table 3.6.

The buffered voltage V_{out1} is attenuated to $V_{in2} = \alpha V_{out1}$ through the voltage divider formed by resistors R_1 and R_2, with α given by

$$\alpha = \frac{R_2}{R_1 + R_2} \qquad (3.13)$$

The attenuated voltage αV_{in} is the input of the main V-I converter formed by OTA-T_1-T_{1C} and resistor R_S. Thus, αV_{in} is followed to V_A, according to

$$\left.\frac{V_A}{V_{in2}}\right|_{\omega \to 0} = \frac{g_{m1}(r_{01}||R_S)GR}{1 + GRg_{m1}(r_{01}||R_S)} \approx 1 \qquad (3.14)$$

where G and R are also given in Table 3.6.

Thereby, the transconductance G_M of the FFVA V-I converter can be approximated as

$$\left.G_M\right|_{\omega \to 0} = \frac{I_{out}}{V_{in}} \approx \alpha \frac{(W/L)_2}{(W/L)_1} \frac{1}{R_S} \qquad (3.15)$$

As for the frequency response, considering the Miller compensated dominant pole approximation from (3.5), the voltage V_A across resistor R_S is

$$\frac{V_A}{V_{in}}(s) \approx \alpha \frac{GBW}{(GBW + s)} \frac{GBW_A}{(GBW_A + s)} \qquad (3.16)$$

and the transconductance G_M is therefore given by

$$G_M(s) = \frac{\alpha}{R_S} \frac{(W/L)_2}{(W/L)_1} \frac{GBW}{(GBW + s)} \frac{GBW_A}{(GBW_A + s)} \qquad (3.17)$$

being $GBW = g_{m1}GR_S/C'_T$; $GBW_A = g_{m1\_A}G_{\_A}(R_1{+}R_2)/C'_{T\_A}$; $C'_T = C + C_{gs1} + C_{gs2} + C_C$ and $C'_{T\_A} = C_{\_A} + C_{gs1\_A} + C_{C\_A}$. This converter exhibits a reduction in bandwidth typical of two cascaded amplifiers.

This G_M must be temperature-independent. Analyzing the expression for the output current, it can be seen that the critical element that makes the system vary with temperature is the resistor R_S. Therefore, as in the previous approach, the immunization of the resistor against temperature variations is made by implementing R_S as the serial connection of two resistors with opposite

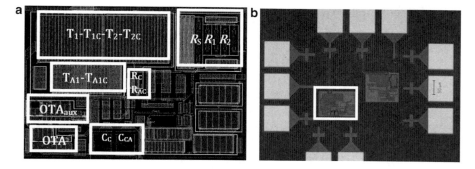

Fig. 3.15 (a) Layout and (b) microphotograph of the FFVA

temperature coefficients, a HRP resistor R_N and a PND resistor R_P resistor, with a ratio $\beta = R_{0N}/R_{0P} = 1.5$. The involved thermal coefficients, which fulfill the equation (3.11), are given in Table 3.3. Note that resistors in the divider, R_1 and R_2, do not need to be temperature compensated, neither to have accurate values because, as long as they are implemented with the same layer and well matched, their ratio will remain constant. Therefore, they are implemented with a HRP layer to minimize the area.

The input range of OTA_{aux} swings from 0 to V_{DD}. Therefore, the use of a rail-to-rail OTA as the one shown in Fig. 3.13 is needed. The resistor that makes the V-I conversion is set to $R_S = 40\,k\Omega$ ($R_{HRP} = 24\,k\Omega$ and $R_{PND} = 16\,k\Omega$). Resistors in the voltage divider are set to $R_{1,HRP} = R_{2,HRP} = 24\,k\Omega$, so that the attenuation factor is $\alpha = 0.5$. As a result, the input common mode voltage for the main OTA, displayed in grey in Fig. 3.14, ranges from 0 to $V_{in}/2$. Hence not a rail-to-rail but a simple PMOS input stage OTA is used, obtained removing transistors M_1, M_2, M_3 and M_6 from the OTA shown in Fig. 3.13. All resistors are implemented with twice the minimum width to reduce deviations from the nominal values. In addition, R_1 and R_2 are implemented with an interdigitation technique for getting a good matching.

Transistors T_1–T_{1C}–T_2–T_{2C} are dimensioned to (50 μm/1.2 μm), so that the scaling factor K between T_1 and T_2 is set to $K = 1$. Sizes of transistors T_{A1}–T_{A1C} are (50 μm/0.25 μm). To have a good matching between M_9–M_{10}, M_{A1}–M_{A2}, and M_{A9}–M_{A10} they are integrated using common-centroid layout techniques including also dummy transistors. The compensation MIM capacitors are $C_C = C_A = 0.4\,pF$. The compensation resistor R_C is the parallel connection of a NMOS (2 μm/0.5 μm) and a PMOS (4 μm/0.5 μm) transistors with $V_{GN} = V_{DD}$ and $V_{GP} = 0\,V$. The biasing current is set to $2I_B = 0.5\,μA$ and it is introduced into the circuit through a current mirror formed by transistors M_B with sizes 10 μm/1.2 μm in the case of the NMOS pair and 20 μm/1.2 μm in the case of the PMOS pair.

The layout and the die microphotograph of the FFVA are shown in Fig. 3.15.

Figure 3.16a shows, at room temperature, the FFVA DC transfer response, measured as the output voltage V_{out} across an external load resistor of $R_L = 40\,k\Omega$ vs. the input voltage V_{in}, compared to the ideal behavior. Figure 3.16b shows the

Fig. 3.16 FFVA response at room temperature: (**a**) output voltage vs. input voltage and (**b**) normalized transconductance

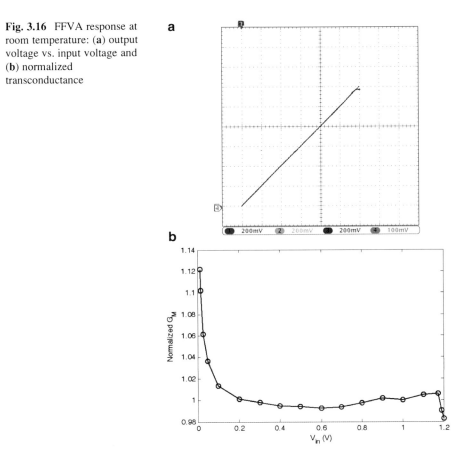

normalized transconductance G_M of the system. The FFVA shows a maximum G_M deviation of 12 %, achieving high linearity in the V-I conversion (0.015 % linearity error) over an input range of (0.0, 1.19 V).

The evaluation of the system against temperature variations, measured with the same equipment than in the previous converters, is next considered. Figure 3.17 shows the output current I_{out} vs. the input voltage V_{in} at different temperatures together with the maximum relative error, corresponding to $T = +120\ °C$. The transconductance of the system varies ±1.0 % over a $(-40, +120\ °C)$ range, with respect to the one at room temperature.

The main performances of the FFVA are summarized in Table 3.7. It exhibits, with power consumption below 75 μW, high linearity (0.015 % linearity error at room temperature) over a (0.0, 1.19 V) input range. It does not show a full input range due to transistor T_{A1} with a small W/L ratio; as explained for the enhanced V-I converter, to keep the bandwidth high (5.2 MHz), this ratio is kept small, thus compromising the input range.

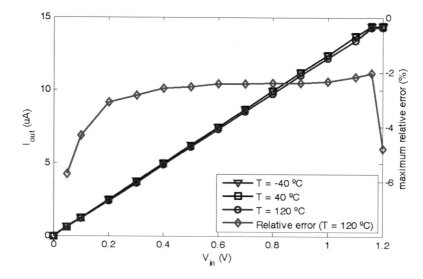

Fig. 3.17 Output current vs. input voltage of the FFVA vs. temperature variation from −40 to +120 °C and the maximum relative error corresponding to $T = +120$ °C

Table 3.7 Measured main performances of the FFVA with a load resistance of $R_L = 40$ kΩ

Parameter	Value
Technology	0.18 μm CMOS
Supply voltage	1.2 V
Transconductance	12.5 μS
G_M deviation	12.2 %
G_M deviation (−40, +120 °C)	30 %
Input range	0.00 – 1.19 V
BW	5.2 MHz
THD (1 V_{pp}@100 kHz)	−42.0 dB
Input offset voltage	22.7 mV
Power consumption	75 μW

3.1.2.2 FBVA: Feedback Voltage Attenuation V-I Converter

The feedback voltage attenuation (FBVA) V-I converter approach is shown in Fig. 3.18. In this structure, to reduce the voltage in node A the floating dynamic battery introduced between the OTA non-inverting input and node A is implemented by using a non-inverting amplifier formed by OTA_{aux}–T_{A1}–T_{A1C} and feedback resistors R_1 and R_2.

This circuit works as follows: the positive input of the main OTA is $V_{\text{in}} = V_{\text{out1}}$. Then, the voltage at node A is $V_A = \alpha V_{\text{in}}$, being α given by

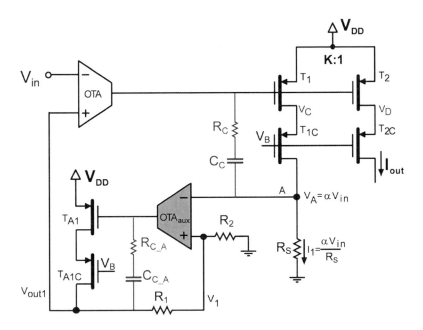

Fig. 3.18 Feedback voltage attenuation V-I converter (FBVA)

$$\alpha = \frac{R_2}{R_1 + R_2} \tag{3.18}$$

This attenuated voltage V_A is converted into a current $I_1 = V_A/R_S$ through resistor R_S. The generated current is then replicated through transistors T_2–T_{2C} with a scaling factor $K{:}1$, resulting an output current $I_{out} = \alpha V_{in}/KR_S$.

Thus, the buffered voltage V_A across R_S is given by

$$\left.\frac{V_A}{V_{in}}\right|_{\omega \to 0} = \frac{(r_{01}\|R_S)g_{m1}GR}{1 + (r_{01}\|R_S)g_{m1}GR\frac{1}{\alpha}} \approx \alpha \tag{3.19}$$

and the transconductance is expressed as follows:

$$\left.G_M\right|_{\omega \to 0} = \frac{g_{m2}GR}{1 + (r_{01}\|R_S)g_{m1}GR\frac{1}{\alpha}} \approx \frac{\alpha}{R_S}\frac{g_{m2}}{g_{m1}} = \frac{\alpha}{R_S}\frac{(W/L)_2}{(W/L)_1} \tag{3.20}$$

To get the frequency response of the system, compensation networks (shown in grey in Fig. 3.18) are included in both OTA and OTA_{aux}. Thus, the attenuated voltage V_A can be approximated by

$$\frac{V_A}{V_{in}}(s) \approx \frac{GBW(\alpha GBW_A + s)}{s^2 + \alpha GBW_A s + GBW_A GBW} \tag{3.21}$$

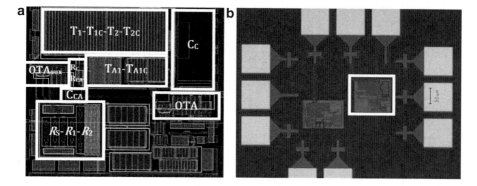

Fig. 3.19 (a) Layout and (b) microphotograph of the FBVA

where the subscrip $_A$ is used to refer to auxiliary elements OTA_{aux}, T_{A1}, T_{A1C}, C_{C_A} and R_{C_A}. and the transconductance of the FBVA is given by

$$G_M = \frac{I_{out}}{V_{in}} \approx \frac{(W/L)_2}{(W/L)_1} \frac{1}{R_S} \frac{GBW(\alpha GBW_A + s)}{s^2 + \alpha GBW_A s + GBW_A GBW} \qquad (3.22)$$

where $GBW = G g_{m1}(r_{01}\|R_S)/C'_T$; $GBW_A = G_{\_A}g_{m1\_A}(r_{0\_A}\|(R_1+R_2))/C'_{T\_A}$; and α is given by (3.18). The response of this converters exhibits one zero and two poles. Hence, an extension on the system bandwidth can be achieved if a pole-zero cancellation is carried out.

Analyzing G_M, its dependence with temperature is due to R_S. Therefore, again this resistor is implemented by the serial connection of R_N and R_P made with HRP and PND layers (thermal coefficients in Table 3.3), being $\beta = R_{0N}/R_{0P} = 1.5$.

Note that, as occur in the FFVA, resistors R_1 and R_2 determine the ratio α and hence do not need to be temperature compensated neither have accurate specified values. Therefore, R_1 and R_2 are implemented using a HRP layer to optimize area.

The input range of the main OTA swings from 0 to V_{DD}, therefore, a rail-to-rail OTA is used. The composite resistor is $R_S = 40\,k\Omega$ ($R_{HRP} = 24\,k\Omega + R_{PND} = 16\,k\Omega$). Feedback resistors are $R_{1,HRP} = R_{2,HRP} = 24$ kΩ, so that the attenuation factor is $\alpha = 0.5$. Therefore, not a rail-to-rail but a simple PMOS input stage OTA is used for the OTA_{aux}. Transistor T_1–T_{1C}–T_2–T_{2C} dimensions are (50 μm/1.2 μm), so the scaling factor is $K = 1$ and transistors T_{A1}–T_{A1C} are set to (50 μm/0.25 μm). In the implementation of this converter, the same layout techniques than those used in the FFVA are used. The MIM compensation capacitors are $C_C = 1.8$ pF and $C_{C\_A} = 3$ fF, and the compensation resistors R_C and R_{C_A} are made with a NMOS (0.3 μm/1.2 μm) and a PMOS (0.6 μm/1.2 μm) in parallel. The biasing current is $2I_B = 0.5$ μA.

Figure 3.19 shows the detailed layout and the die microphotograph of the FBVA V-I converter. Figure 3.20a shows, at room temperature, the FBVA DC transfer response, measured as the output voltage V_{out} across an external load resistor

Fig. 3.20 FBVA response at
room temperature: (**a**) output
voltage vs. input voltage and
(**b**) normalized
transconductance

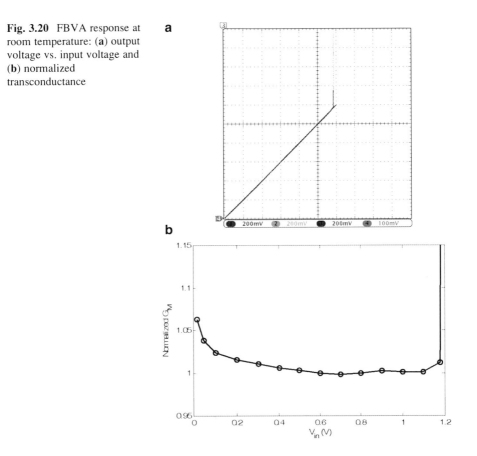

$R_L = 40 \text{ k}\Omega$ vs. the input voltage V_{in}, compared to the ideal behavior. Figure 3.20b
shows the normalized G_M of the system with $R_L = 40 \text{ k}\Omega$. It shows a maximum G_M
deviation of 6.3 %, achieving a linearity error in the V-I conversion of 0.014 % over
an input range from 0 up to 1.18 V.

For a -40 to $+120$ °C range, the measured output current I_{out} vs. the input
voltage V_{in} at several temperatures and the maximum relative error corresponding
to $T = -40$ °C are shown in Fig. 3.21. The transconductance of the system varies
± 1.4 % over a $(-40, +120$ °C$)$ range, with respect to the one at room temperature,
with a maximum relative error of 8.5 %.

Table 3.8 summarizes the main performances of the FBVA. It has an input range
of $(0.0, 1.18 \text{ V})$, slightly limited by the $(W/L)_{1\_A}$ ratio due to an optimization of the
bandwidth (14.5 MHz). The power consumption is 70 µW.

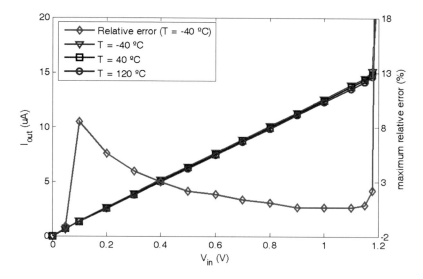

Fig. 3.21 Output current vs. input voltage of the FBVA at different temperatures and the maximum relative error corresponding to $T = -40\,°C$

Table 3.8 Measured main performances of the FBVA with a load resistance of $R_L = 40\ k\Omega$

Parameter	Value
Technology	0.18 µm CMOS
Supply voltage	1.2 V
Transconductance	12.5 µS
G_M deviation	6.3 %
G_M deviation (−40, +120 °C)	8.5 %
Input range	0.00 – 1.18 V
BW	14.5 MHz
THD (1 V_{pp}@100 kHz)	−51.6 dB
Input offset voltage	20.6 mV
Power consumption	70 µW

3.1.2.3   FFCA: Feedforward Current Attenuation V-I Converter

The last rail-to-rail V-I converter is the feedforward current attenuation (FFCA) V-I converter. In this approach, illustrated in Fig. 3.22, the floating dynamic battery between the OTA non-inverting input and node A is implemented using a linear resistor R_2 driven by a current source proportional to V_{in}. To generate the required current source, an auxiliary OTA V-I converter formed by OTA_{aux}, T_{A1}–T_{A1C} and R_1 is used.

Considering the subscript $_A$ to refer to auxiliary elements, the input voltage V_{in} is buffered to V_{out1} by the OTA_{aux}–T_{A1}–T_{A1C} voltage follower with rail-to-rail input–output voltage operation, as given by

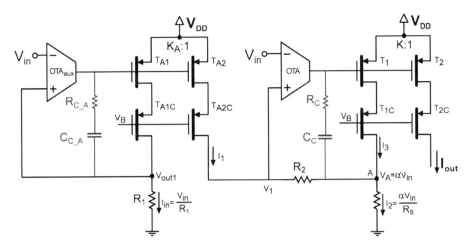

Fig. 3.22 Feedforward current attenuation V-I converter (FFCA)

$$\left.\frac{V_{out1}}{V_{in}}\right|_{\omega\to0} = \frac{(g_{m1\_A}G_{\_A})(r_{01\_A}||R_1)R_{\_A}}{1 + g_{m1\_A}G_{\_A}(r_{01\_A}||R_1)R_{\_A}} \approx 1 \qquad (3.23)$$

The generated current I_{in} across resistor R_1 is replicated through transistors T_{A1}-T_{A1C}-T_{A2}-T_{A2C} with a scaling factor K_A:1, $K_A = (W/L)_{1\_A}/(W/L)_{2\_A}$. Thus, according to (3.4), the transconductance for the current I_1 is given by

$$\left.\frac{I_1}{V_{in}}\right|_{\omega\to0} = \frac{(g_{m2\_A}G_{\_A})R_{\_A}}{1 + G_{\_A}R_{\_A}(r_{01\_A}||R_1)g_{m1\_A}} \approx \frac{g_{m2\_A}}{g_{m1\_A}}\frac{1}{R_1} = \frac{(W/L)_{2\_A}}{(W/L)_{1\_A}}\frac{1}{R_1}$$
$$= \frac{1}{K_A R_1} \qquad (3.24)$$

The current I_1 drives a resistor R_2, so that the voltage across R_2 is $V_{R2} = I_1 R_2$. The positive input of the main OTA is $V_1 = V_A + I_1 R_2$, whereas the negative input is at V_{in}. A straight analysis shows that, $V_1 = V_{in}$, so $V_A = V_1 - I_1 R_2 = V_{in} - V_{in}/K_A R_1 = \alpha V_{in}$, being α is the attenuation factor, which fulfills the following expression:

$$\alpha = \frac{K_A R_1 - R_2}{K_A R_1} \qquad (3.25)$$

If the analysis is carried out with small signal models, the voltage V_A is given by

$$\left.\frac{V_A}{V_{in}}\right|_{\omega\to0} = \frac{g_{m1}(r_{01}||R_S)RG\alpha}{1 + g_{m1}(r_{01}||R_S)RG} \approx \alpha \qquad (3.26)$$

validating that the voltage in node A is attenuated.

The output current I_{out} is the replica of I_3, being $I_3 = I_1 - I_2$ and can be approximated as $I_{out} = I_3(W/L)_2/(W/L)_1$. Thus, the transconductance G_M of the FFCA will be given by

$$G_M\big|_{\omega \to 0} = \frac{(W/L)_2}{(W/L)_1} \frac{K_A R_1 - R_2 - R_S}{K_A R_1 R_S} \qquad (3.27)$$

The frequency response of the converter, assuming a dominant pole compensation with $C'_T = C + C_C + C_{gs1} + C_{gs2}$ and $C'_{T\_A} = C_{\_A} + C_{C\_A} + C_{gs1\_A} + C_{gs2\_A}$, gives an attenuated voltage V_A

$$\frac{V_A}{V_{in}}(s) = \frac{GBW(\alpha GBW_A + s)}{(GBW_A + s)(GBW + s)} \qquad (3.28)$$

And the transconductance of the system is given by

$$G_M(s) = \frac{(W/L)_2}{(W/L)_1} \frac{GBW GBW_A (K_A R_1 \alpha - R_S) + (K_A R_1 GBW - R_S GBW_A)s}{R_S K_A R_1 (GBW_A + s)(GBW + s)} \qquad (3.29)$$

where $GBW = g_{m1} G R_S / C'_T$ and $GBW_A = g_{m1\_A} G_{\_A} R_1 / C'_{T\_A}$.

The transfer function presents one zero and two poles, and if a pole-zero cancellation is achieved, a bandwidth broadening is obtained. Note that in the auxiliary V-I converter the current copy across transistors $T_{A1}-T_{A2}$ is not seriously degraded when they enter the triode region, since $T_{A1}-T_{A2}$ maintain the same gate and drain voltage values over the whole operating range. Remember that to keep the current sensing transistors T_1-T_{1C} working in the saturation region for all the input voltage, the attenuation factor must be $\alpha < 0.9$. Note that in this converter it also exhibits a lower limit: from (3.27), to achieve a positive current I_{out}, the condition $\alpha > R_S/K_A R_1$ must be satisfied. Therefore, overall, $R_S/K_A R_1 < \alpha < 0.9$ has to be fulfilled.

The G_M dependence with temperature is due to resistor R_S, which performs the main V-I conversion, as well as to resistors R_1 and R_2 which are introduced to attain the voltage reduction. Therefore, composite resistors are used for all R_S, R_1, R_2, made with $R_N = R_{HRP}$ and $R_P = R_{PND}$ (thermal coefficients in Table 3.3), being $\beta = R_{ON}/R_{OP} = 1.5$. These resistors are set to $R_S = R_1 = R_2 = 40\,k\Omega$ ($R_{HRP} = 24\,k\Omega$ and $R_{PND} = 16\,k\Omega$).

The input range of both the OTA and the OTA$_{aux}$ is going to swing from 0 to V_{DD}, therefore, OTAs with two complementary differential pairs are used. Transistors $T_1-T_{1C}-T_2-T_{2C}$ dimensions are set to (50 µm/1.2 µm), so $K = 1$. Transistors $T_{A1}-T_{A1C}$ are set to (50 µm/1.2 µm) and $T_{A2}-T_{A2C}$ to (12.5 µm/ 1.2 µm). So $K_A = 4$, obtaining an attenuation factor $\alpha = 0.75$. The MIM compensation capacitors are $C_C = 550$ fF and $C_{C\_A} = 285$ fF, and the compensation resistors are the parallel connection of a NMOS (0.5 µm/1.0 µm) and a PMOS (1.0 µm/1.0 µm) in the main OTA and a NMOS (0.5 µm/1.2 µm) and a PMOS (1.0 µm/1.2 µm) in OTA$_{aux}$. The biasing current is set to $2I_B = 0.5$ µA.

The layout and the die photograph of the FFCA V-I converter is shown in Fig. 3.23. Figure 3.24a shows, at room temperature, the FFCA DC transfer response, measured as the output voltage V_{out} across an external load resistor $R_L = 40$ kΩ versus the input voltage V_{in}, compared to the ideal behavior.

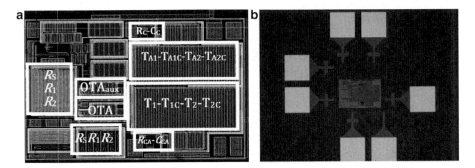

Fig. 3.23 (**a**) Layout and (**b**) microphotograph of the FFCA

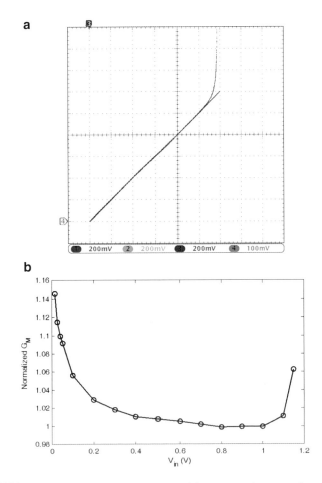

Fig. 3.24 FFCA response at room temperature: (**a**) output voltage vs. input voltage and (**b**) normalized transconductance

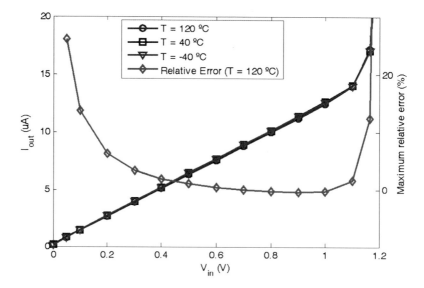

Fig. 3.25 Output current vs. input voltage of the FFCA at different temperatures and the maximum relative error corresponding to $T = +120\,^{\circ}\mathrm{C}$

Table 3.9 Measured main performances of the FFCA with a load resistance of $R_L = 40\,\mathrm{k\Omega}$

Parameter	Value
Technology	0.18 µm CMOS
Supply voltage	1.2 V
Transconductance	12.5 µS
G_M deviation	14.5 %
G_M deviation $(-40, +120\,^{\circ}\mathrm{C})$	26.1 %
Input range	0.0 – 1.10 V
BW	14.1 MHz
THD (1 V_{pp}@100 kHz)	−44.2 dB
Input offset voltage	18.7 mV
Power consumption	80 µW

Figure 3.24b shows the normalized transconductance G_M. The FFCA shows a maximum G_M deviation of 14.5 %, achieving a linearity error in the V-I conversion of 0.006 % over an input range from 0 up to 1.10 V.

For a −40 to +120 °C range, the measured output current I_{out} vs. the input voltage V_{in} at different temperatures and the maximum relative error, corresponding to T = +120 °C are shown in Fig. 3.25b. The transconductance of the system varies ±1.0 % over a (−40, +120 °C), with respect to the one at room temperature, with a maximum relative error of 26.1 %.

Table 3.9 summarizes the main performances of the FFCA. It has an input range of (0.0, 1.10 V), slightly limited by the $(W/L)_{1\_A}$ ratio due to an optimization of the bandwidth (14.1 MHz). However, in this case, if the $(W/L)_{1\_A}$ is increased, there will still exist a slight reduction on the input range due to transistors T_{1A} and T_{2A} entering in the triode region. The power consumption is 80 μW.

3.1.3 Summary

Figure 3.26a shows the normalized output current I_{out} and Fig. 3.26b the G_M deviation for the three proposed rail-to-rail V-I converters, compared with the conventional and the enhanced V-I converters. In order to make a fair comparison, the conventional and the enhanced converters are also fed with a single supply of 1.2 V and use the rail-to-rail OTA shown in Fig. 3.13. Therefore, note that their results are simulated. The displayed behavior agree with the expected response: the conventional proposal has a rather limited input range (0 – 0.68 V), which is significantly extended, but still limited (0 – 0.94 V) with the enhanced approach, while rail-to-rail operation is achieved with the new FFVA (0 – 1.19 V) and FBVA (0 – 1.18 V) configurations, and almost rail-to-rail (0 – 1.10 V) for the FFCA proposal. The three new converters show a slight reduction on the input range due to the operating range-bandwidth trade-off introduced by the size of the current sensing transistors. If bandwidth is not a relevant design parameter, true rail-to-rail is achieved in the voltage-attenuation configurations (FFVA and FBVA). However, FFCA will present a limit, around 1.17 V due to transistors T_{A1} and T_{A2} entering in triode region for input voltages above 1 V, as it occurs in the enhanced approach. However, thanks to the negative feedback, $V_{DS,T2} = V_{DS,T1}$, allowing a good current copy in an extended range.

To validate the widening in the input range and the narrowing in the bandwidth with T_1 and T_{1A} dimensions, simulations have been carried out for $(W/L)_1$ and $(W/L)_{A1}$ being (50 μm/1.2 μm) and (300 μm/1.2 μm). Figure 3.27 shows the DC transfer for voltages above 1 V for the three converters, and their frequency response, highlighting the compromise between bandwidth and operating range. For the integrated structures, with $(W/L)_1 = (W/L)_{1\_A} = (50$ μm/1.2 μm), the FFVA presents the typical response of two amplifiers in cascade, so it exhibits a reduction on the bandwidth with respect to a single stage approach, being its bandwidth $BW_{FFVA} = 5.2$ MHz. Meanwhile, the transfer function of both the FBVA and the FFCA present one zero and two poles, and if a pole-zero cancellation is achieved, an extension on the bandwidth is obtained. Therefore both exhibit a similar enhanced bandwidth frequency behavior as shown by the results, being $BW_{FBVA} = 14.5$ MHz and $BW_{FFCA} = 14.1$ MHz.

To close the study of V-I converters, a summary of the features of the three considered structures is offered in Table 3.10, compared to some of the rail-to-rail V-I converters found in the literature [HAS11, HUN99, LOP07, VER95]. The proposed structures offer low power operation with rail-to-rail input range and bandwidths that are enough for the environmental application, where slow varying signal will be introduced in the converter. On the whole, a good trade-off between the main characteristics is obtained for all the given structures.

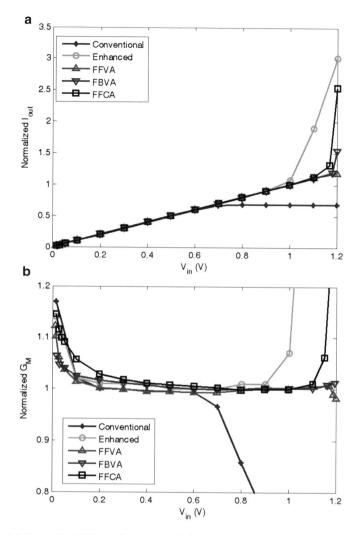

Fig. 3.26 (**a**) Normalized DC transfer characteristic and (**b**) normalized G_M at room temperature, for the simulated conventional and the enhanced V-I converters and the measured FFVA, FBVA and FFCA V-I converters

3.2 Bidirectional Current Integrators

The bidirectional current integrator, also known as the charge/discharge circuit, consists of a timing capacitor C_{int}, a constant current source that charges the capacitor and a constant current sink that discharges it.

To control the capacitor charge and discharge process, the selected option in this work is to introduce, after the integrator, a control circuit that senses the voltage across the timing capacitor, and compares it with an upper (V_H) and a lower (V_L)

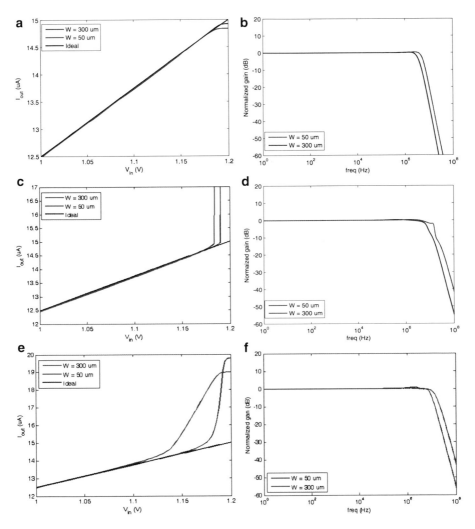

Fig. 3.27 Output current vs. input voltage and frequency response for different T_1–T_{1A} dimensions: (**a**) and (**b**) FFVA, (**c**) and (**d**) FBVA and (**e**, **f**) FFCA

limits. When the capacitor voltage reaches one of those limits the control circuit acts in such a way that it reverses the charging phase.

Let us assume that the capacitor is initially discharged, and the output of the control circuit is such that a current source I is connected to the capacitor. The capacitor is charged linearly with a constant current, being the behavior of the voltage across capacitor V_{cap} given by $V_{cap}(t) = V_0 + It/C_{int}$, where V_0 is the voltage across capacitor at $t = 0$. When V_{cap} reaches the upper limit V_H, the control circuit output changes, so that a current sink I is now connected to the capacitor, starting to linearly discharge it. When V_{cap} reaches the lower limit V_L, the charging phase starts again. The charge/discharge phases are repeated, being the circuit in

Table 3.10 Comparison of rail-to-rail V-I converters

Parameter	[VER95]	[HUN99]	[LOP07]	[HAS11]	FFVA	FBVA	FFCA
Technology	2.5 μm CMOS	1.2 μm CMOS	0.5 μm CMOS	0.18 μm CMOS	0.18 μm CMOS	0.18 μm CMOS	0.18 μm CMOS
Supply voltage (V)	3	3	±1.5	1	1.2	1.2	1.2
G_M (μS)	116	20	10	100	12.5	12.5	12.5
G_M deviation (T = 27 °C) (%)	3[a]	13[a]	–	3[b]	12.2[a]	6.3[a]	14.5[a]
G_M deviation: −40 to +120 °C (%)	–	–	–	–	30[a]	8.5[a]	26.1[a]
Input range (V)	0 – 3	0 – 2.8	0 – 3	0 – 1	0 – 1.2	0 – 1.2	0 – 1.17
Bandwidth (MHz)	4.8	–	90	39.2	5.2	14.1	14.5
THD	−57.08 dB (0.4V_{pp}@ 1 kHz)	−37.4 dB (1V_{pp} @1 kHz)	−60 dB (6V_{pp}@ 100 kHz)	−44.4 dB (1V_{pp} @1 MHz)	−42.0 dB (1V_{pp}@ 100 kHz)	−51.6 dB (1V_{pp}@ 100 kHz)	−44.2 dB (1V_{pp}@ 100 kHz)
Power (μW)	1,200	310	3,000	730	75	75	80
Area (mm²)	–	0.129	0.1	–	0.0102	0.0112	0.0144

[a]Full input range
[b]$V_{in} > 0.1$ V

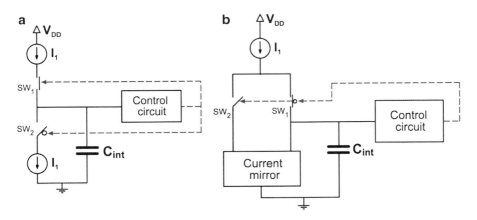

Fig. 3.28 Bidirectional current integrators with: (**a**) complementary current sources and (**b**) a current source and a current mirror

astable operation. The time required to charge/discharge the capacitor between the limits V_H and V_L is given by

$$t = \frac{T}{2} = \frac{C_{int}(V_H - V_L)}{I}$$

(3.30)

being $T/2$ the half period of the generated signals: the one across the capacitor (V_{cap}) and the control circuit output. As the capacitor is being charged/discharged at a constant rate, voltage across the capacitor ideally will be a triangular wave, and the control circuit output a square wave.

In this section, two solutions to implement the bidirectional current integrator will be differentiated: the conventional and a low-power solution. Both, as shown in Fig. 3.28, use a grounded capacitor C_{int} and use the current that comes from the V-I converter as currents I_1 that charge and discharge the capacitor. The grounded capacitor bidirectional integrator shows structural advantages of high frequency and high temperature performances compared with those with a floating capacitor in an OTA feedback loop, while reduces the required power consumption. In addition, grounded capacitors are usually more linear than floating ones [CAI95].

Regarding the temperature dependence of the capacitor, it is going to be implemented with a MIM capacitor from the UMC 0.18-µm technology. Therefore, the temperature drift given by the foundry is 40 ppm/°C. In the temperature range in which the VFC is going to work, $(-40, +120\,°C)$, the drift in the capacitor value is less than $\pm 0.4\,\%$, thus, the temperature drift due to the capacitor will be neglected since it will not be a dominant source of error. However, the MIM capacitors are sensitive to the dielectric thickness variation and the metal plates geometry. Typically, a modern CMOS process can maintain a variation in some integrated capacitors of $\pm 20\,\%$ [CHE02]. However, in [LIM04], a comparison of 165 samples of MIM capacitors is carried out obtaining a maximum standard deviation of only 0.11 %.

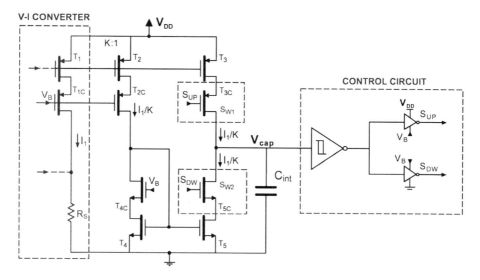

Fig. 3.29 Implementation of the conventional bidirectional current integrator

3.2.1 Conventional Bidirectional Current Integrator

The conceptual scheme of the conventional current integrator is shown in Fig. 3.28a. Assuming initially that the switch SW_1 is closed and SW_2 is open, the timing capacitor C_{int} is charged with a current I_1 until it reaches the upper limit voltage V_H. In that moment, the control circuit causes that the switches SW_1 and SW_2 to change their states. Then, the current I_1 discharges the capacitor until it reaches the lower limit V_L, starting next the charge phase over again.

Practical implementation of this integrator is shown in Fig. 3.29. The current I_1, that has been generated in the V-I converter, is replicated with a scaling factor K:1 twice: the current replicated through transistors T_3–T_{3C} constitutes the charging current, whereas the current replicated through transistors T_2–T_{2C} is mirrored through the high-swing cascode mirror formed by transistors T_4–T_{4C} and T_5–T_{5C}, providing the discharging current. Cascode transistors T_{3C} and T_{5C} constitute the switches SW_1 and SW_2 respectively, and they are controlled by means of their gate voltage V_B, which has been chosen to be the same for PMOS and NMOS transistors for simplicity.

During the charge phase, the output of the control circuit is '1', $S_{UP} = V_B$ and $S_{DW} = 0$ V. Thus, T_{3C} is ON whereas T_{5C} is OFF. When the voltage across capacitor C_{int} reaches the upper limit V_H, the control circuit output changes to '0,' then $S_{UP} = V_{DD}$ and $S_{DW} = V_B$, therefore, T_{3C} is OFF and T_{5C} is ON, starting the discharging phase. When the voltage across capacitor C_{int} reaches the lower limit, the control circuit output changes to '1', then $S_{UP} = V_B$ and $S_{DW} = 0$ V; therefore, T_{3C} is ON and T_{5C} goes to OFF, starting again the charging phase.

Thus, cascode transistors are used to improve the current copy. Besides, they act as switching elements, achieving a high performance compact topology.

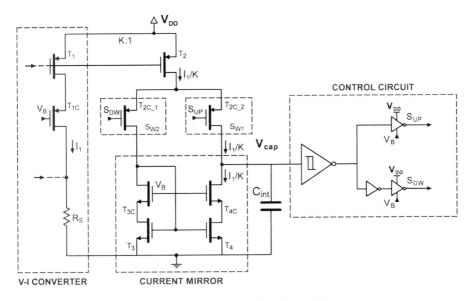

Fig. 3.30 Implementation of the low-power bidirectional current integrator

3.2.2   Low-Power Bidirectional Current Integrator

The general scheme of the low-power current integrator is shown in Fig. 3.28b. Assuming that the switch SW_1 is closed and SW_2 is open, the timing capacitor C_{int} is charged directly with the current I_1 until it reaches the upper limit V_H. In that moment, the control circuit causes switches SW_1 and SW_2 to change their states. Then, the current I_1 is mirrored and discharges the capacitor until it reaches the lower limit voltage V_L, starting the charging phase over again. This type of integrator is less power demanding because there is only a current source.

Practical implementation of this integrator is shown in Fig. 3.30. The current I_1, that has been generated in the V-I converter, is now replicated with a scaling factor K:1 only once through transistor T_2. In the charging phase, the current flows through T_{2C_2}, whereas in the discharging phase, the current flows through T_{2C_1}, and it is then mirrored through the high-swing cascode mirror formed by transistors T_3–T_{3C} and T_4–T_{4C}. In this case, cascode transistors T_{2C_1} and T_{2C_2} constitute the switches SW_1 and SW_2 respectively, and they are controlled by means of their gate voltages, being V_B the biasing gate voltage.

During the charge phase, the output of the control circuit is '1', $S_{UP} = V_B$ and $S_{DW} = V_{DD}$. Thus, T_{2C_2} is ON whereas T_{2C_1} is OFF. When the voltage across capacitor C_{int} reaches the upper limit V_H, the control circuit output changes to '0,' then $S_{UP} = V_{DD}$ and $S_{DW} = V_B$; therefore, T_{2C_2} is OFF, and T_{2C_1} is ON, starting the discharging phase. During the discharge phase, the current I_1/K is mirrored through the high-swing cascode mirror. When the voltage across capacitor C_{int}

reaches the lower limit V_L, the control circuit output changes to '1,' then $S_{UP} = V_B$ and $S_{DW} = V_{DD}$, starting again the charging phase.

Just as for the conventional case, the cascode configuration provides an accurate current copy and avoids the use of additional switches.

3.3 Control Circuits

The control circuit is the responsible for providing the digital control signals SW_1, SW_2 that will control the direction of the current in the bidirectional integrator. Voltage window comparators (VWC) or Schmitt triggers are the most common choices used as the control circuit.

A Schmitt trigger comparator, as well as other comparators with an inherent hysteresis, is not suitable to be used in these systems because their limits V_H and V_L will strongly depend on temperature or supply voltage variations. Due to this limitation, the preferred alternative is the use of a voltage-window comparator, whose block diagram is shown in Fig. 3.31a. It consists of two comparators and $\overline{RS}$ flip-flop that provides stable control signals. Each of the comparators compares the input signal V_{cap} with one of the limits, V_H or V_L. The $\overline{RS}$ flip-flop changes its output when one of the comparator output changes, that is, when V_{cap} reaches either V_H or V_L, providing stable output control signals. The operation of the VWC is shown in Table 3.11.

When the capacitor is being charged, as $V_L < V_{cap} < V_H$, both comparators have a '1' at their output, therefore, according to Table 3.11, Q and $\bar{Q}$ remains in their previous states. At the moment that V_{cap} reaches V_H, the output of the upper comparator changes to '0', $Q = $ '0' and $\bar{Q} = $ '1'. In that moment, the discharging phase starts, and again $V_L < V_{cap} < V_H$, keeping Q and $\bar{Q}$ in their previous states, that is, $Q = $ '1' and $\bar{Q} = $ '1'. When V_{cap} reaches V_L, the output of the lower comparator changes to '0', thus being $Q = $ '1' and $\bar{Q} = $ '0'. At that instant the

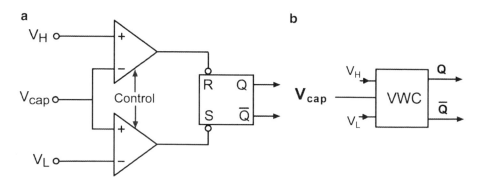

Fig. 3.31 (a) Block diagram and (b) symbol of a VWC

Table 3.11 Operation of a VWC

V_{cap}	$\bar{S}$	$\bar{R}$	Q	$\bar{Q}$
–	0	0	0	0
$V_{cap} \leq V_L$	0	1	1	0
$V_{cap} \geq V_H$	1	0	0	1
$V_L < V_{cap} < V_H$	1	1	Last Q	Last $\bar{Q}$

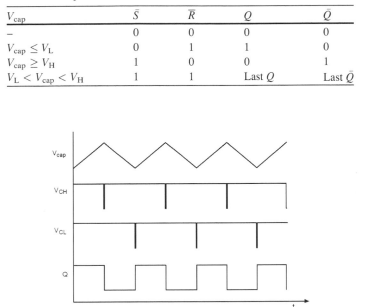

Fig. 3.32 Time-domain response of V_{cap}, V_{CH}, V_{CL} and Q

charging phase starts. Figure 3.32 shows the temporal behavior of the voltage across capacitor, the output of both comparators and the output of the $\overline{RS}$ flip-flop.

An important requirement is speed: it is desirable that the comparators do not cause a delay comparable to the frequencies of interest. This makes the VWC be the most power demanding block of the whole VFC. In order to obtain a fast response, simple structure of comparator circuits must be used. In addition, since power consumption is a key design parameter, only one comparator will be active at a time. That is, when the capacitor is being charged, the only comparator that works is the one that compares the signal with V_H, while when the capacitor is being discharged the only capacitor that is active is the one that compares the signal with V_L.

The relative output frequency error of the bidirectional current integrators and this control circuit is given by (3.31), where T_S is the delay introduced by the VWC and the switches SW_1 and SW_2 [CAI95].

$$\frac{\delta f}{f_0} = -2f_0 T_S \tag{3.31}$$

Note that the output of the control circuit is also the output of the VFC, whose output frequency f_0 is given by

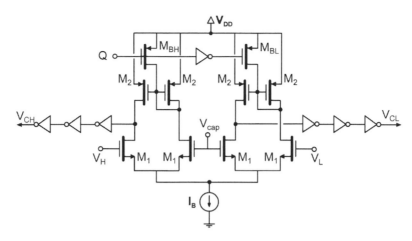

Fig. 3.33 Schematics of the two comparators with their enables

$$f_0 = \frac{I}{2KC_{int}(V_H - V_L)} \tag{3.32}$$

where I is the generated current in the V-I converter and K the scaling factor between current mirrors in the bidirectional current integrator.

3.3.1 Low-Power Comparator

Each comparator is made up of a simple differential pair followed by NOT gates. Both comparators, combined with their enable circuitry (M_{BH}, M_{BL}), are shown in Fig. 3.33. Transistors in the differential pair have minimum length, to decrease the time response of the circuit while increasing the gain. The logic inverters are used to increase the overall comparator gain, and thus increasing the speed. The gain of the stage is approximately given by

$$A_V \approx \frac{-(g_{mP} + g_{mN})^3}{(r_{ON} + r_{OP})^3} \frac{g_{m1}}{(r_{01} + r_{02})} \tag{3.33}$$

where the parameters have their usual meaning, the subscript $_i$ refer to the corresponding transistor M_i and subscript $_P$ and $_N$ refer to transistors in the logic inverters.

Regarding the power reduction, it is achieved by transistors M_{BH} and M_{BL} and the output control signals Q and $\bar{Q}$. When the capacitor is being charged ($Q = $ '1'), the only comparator that is obliged to be active is the upper one. Thus, M_{BH} is OFF, and M_{BL} is ON, deactivating the lower comparator. On the other hand, when the capacitor is being discharged ($Q = $ '0'), being the lower comparator the only one that has to be active. Thus, M_{BH} is ON, and M_{BL} is OFF, deactivating the upper comparator.

3.3.2 Comparison Limits Generation

It is desirable that the threshold voltages V_H and V_L will be compensated against temperature and supply voltage variations, although it is more important to have the difference between them, V_H-V_L, immunized, because the output frequency of the VFC directly depends on this difference.

The easiest way to generate these voltage limits is through a voltage divider made either with passive or active resistors, such as MOS transistors in diode connection. However, in this way, the generated voltages will strongly depend on supply voltage (and probably on temperature) variations.

Another way to generate these voltages is by means of a current I_B flowing across two resistors. If this current has dependences on supply voltage or temperature, the resistors ought to compensate those drifts. Although CMOS technologies usually provide resistors with different temperature coefficients and thus a temperature compensation technique can be implemented, their variation with the supply voltage is negligible, which makes a non trivial task compensate the voltage supply dependence.

Therefore, the suitable option will be to generate a current I_B by a temperature and supply voltage independent current source, and to obtain the comparison limits with $V_{DD}-$ and T-independent resistors, being therefore V_H and V_L immunized against $V_{DD}-$ and T-variations. This solution is explored in more detail in the next section.

3.4 Bias Circuit

The last main block that needs to be study is the bias current generation circuit. The generated current I_B is desired to be immunized against temperature and power supply variations to ensure proper operation for all the blocks in the VFC. There are several easy ways to generate a bias current, for instance with a resistor and a MOS transistor in diode connection, or controlling the gate voltage of a transistor in saturation, but they present important shortcomings: the generated current directly depends on the supply voltage and it does strongly depend on temperature. Therefore, a β-multiplier will be used to generate a reference current.

3.4.1 β-Multiplier

The proposed bias circuit uses a conventional β-multiplier referenced self-biasing circuit [BAK98], as shown in Fig. 3.34. The width of transistor M_2 is made α times larger than the width of transistors M_1. Gate-source voltages of transistors M_1 and M_2 satisfy the relationship:

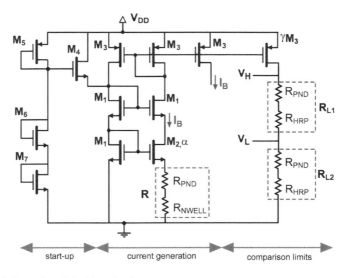

Fig. 3.34 Schematics of the bias circuit

$$V_{GS1} = V_{GS2} + I_B R \tag{3.34}$$

These transistors work in the subthreshold region; therefore, the gate-source voltage also satisfies the following:

$$V_{GSi} = nV_T ln \left(\frac{I_B}{I_{DO}} \right) + V_{th} \tag{3.35}$$

where n is the subthreshold slope factor, V_T is the thermal voltage (26 mV at room temperature) and I_{DO} is the off-drain current. From (3.34) and (3.35) the expression for the bias current I_B can be obtained:

$$I_B = \frac{nV_T}{R} ln(\alpha) \tag{3.36}$$

Cascode transistors are used to ensure the same current I_B flowing through paths. The generated current I_B, as occurs in all self-biased circuits, has two stable states, the one where is zero and the one where I_B is generated. To avoid the possibility that the current was zero, a start-up circuit should be used. Thus, the conventional start-up circuit made up of transistors M_4–M_7 is introduced. If gate voltage of transistor T_4 is at or near 0 V, M_4 turns on and pulls this node upward the corresponding stable state, increasing the gate voltage until it reaches $2V_{GS}$. This causes M_4 to turn off, thus, the start-up circuit does not affect the operation of the bias circuit.

Note that, as shows (3.36), this current is, in first order, power supply independent, but, it strongly depends on temperature because of the thermal voltage

Table 3.12 Thermal coefficients of the used resistors

	$TC_1 (\cdot 10^{-3\circ}C^{-1})$	$TC_2 (\cdot 10^{-7\circ}C^{-2})$
R_{PND}	1.184	7.310
R_{NWELL}	2.504	85.66
R	1.791	43.55
R_{HRP}	-0.834	13.0
R_L	-0.0268	9.08

$(TC_1 = 3 \times 10^{-3\circ}C^{-1})$ [BAK98] and the subthreshold slope factor $(TC_1 \sim -10^{-3\circ}C^{-1})$ [WAN06b] drifts. To compensate the positive temperature coefficient of the output current, the resistor R is chosen such that it has a similar temperature coefficient, so that the fractional sensitivities tend to cancel. Thus, according to the resistors provided by the used technology, R is going to be implemented by means of the serial connection of a NWELL resistor and a P^+ nonsalicide diffusion resistor (PND) with a resistor ratio $R_{PND}/R_{NWELL} = 20/17$. Temperature coefficients of R_{NWELL}, R_{PND} and the composite resistor R are shown in Table 3.12.

As explained in Sect. 3.3, once generated a V_{DD}– and T-independent current I_B, an easy way to generate supply voltage- and temperature-independent comparison limits V_H and V_L is by means of constant resistors. As shown in Fig. 3.34, to drive the resistors, the current I_B is replicated with a factor γ as a compromise between power consumption and resistor values. Therefore, comparison limits are given by $V_H = \gamma I_B (R_{L1} + R_{L2})$ and $V_L = \gamma I_B R_{L2}$, where both R_{L1} and R_{L2} are temperature-independent resistors made as the serial connection of resistances with opposite temperature coefficients: $R_L = R_{HRP} + R_{PND}$, being the temperature coefficients of R_{HRP}, R_{PND} and R_L resistors shown in Table 3.12, and the ratio that immunizes R_L against T variations is $\beta = 1.5$.

To validate the supply voltage and temperature insensitivity of the bias current, it has been simulated in a 0.18-μm CMOS technology from UMC, setting the bias current to $I_B = 500$ nA. According to expression (3.36), if the resistor R is set to $R = 115.6$ kΩ ($R_{PND} = 52$ kΩ $+ R_{NWELL} = 63.6$ kΩ), the factor α is set to 6.

Figure 3.35a shows the variation of the normalized bias current I_B over a variation in the supply voltage from 1.0 to 1.4 V. The current variation is 4.5 nA/V. Figure 3.35b shows the variation of the normalized I_B over the temperature range $(-40, +120 °C)$. In this case, the current variation is 52.5 pA/°C.

In order to see the improvement in the bias current temperature dependence, Fig. 3.36 shows the generated current against temperature variations simulated with the composite resistor ($R_{NWELL} + R_{PND}$), with an ideal resistor (zero temperature-coefficient), and with a HRP resistor. A HRP resistor would be the choice when optimizing area, but the variation of the current because of its temperature coefficients makes it unsuitable in this circuit.

To generate a T-and V_{DD}-independent comparison limits (V_H and V_L), the generated current I_B is replicated with a scaling factor $\gamma = 8$, and comparison

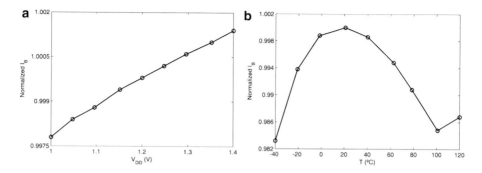

Fig. 3.35 Variation of the normalized generated current I_B against (**a**) supply voltage variations and (**b**) temperature variations

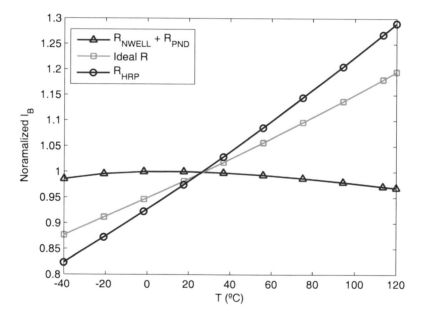

Fig. 3.36 Normalized current generated with an ideal, a HRP, and a composite ($R_{NWELL}+R_{PND}$) resistors

limits are generated by means of resistors $R_{L1} = R_{L2} = 100$ kΩ ($R_{PND} = 40$ kΩ and $R_{HRP} = 60$ kΩ) to get comparison limits $V_L = 0.4$ V and $V_H = 0.8$ V. Figure 3.37a shows the variation of the normalized V_H–V_L over a variation in the supply voltage from 1.0 to 1.4 V, varying 11.3 mV/V. Figure 3.37b shows the variation of the normalized V_H–V_L over the temperature range (-40, $+120$ °C), where it varies 29.4 μV/°C.

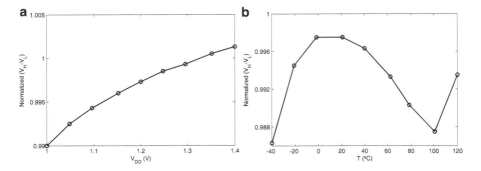

Fig. 3.37 Variation of the normalized generated (V_H-V_L) against (**a**) supply voltage variations and (**b**) temperature variations

3.5   Conclusions

This chapter has introduced the main building blocks necessary to design a multivibrator VFC: V-I converter, bidirectional current integrator, control circuit, and bias circuit.

Rail-to-rail operation is mandatory in low-voltage VFCs to take advantage of the achievable resolution. The input voltage range of the converter is set by the corresponding one of the V-I converter. Therefore, the design of low-voltage low-power rail-to-rail V-I converters with high linearity is mandatory. Targeting this goal, three new different rail-to-rail converters have been proposed based on an OTA/common source amplifier with a floating dynamic battery in the feedback loop which achieve a good trade-off between their main performances compared to the state-of-art converters.

Two bidirectional current integrators suitable to be used in VFC have been presented. However, the low-power bidirectional current integrator presents some advantages: it is less power demanding because the current is only replicated once and therefore charging and discharging current are more alike because of using single current replication.

There are many approaches to obtain a hysteresis comparator, however, in the majority of them the limits V_H and V_L strongly depends on temperature and power supply variations. Therefore, the most suitable option is to use a voltage-window comparator, which is made up of two comparators and a $\overline{RS}$ flip-flop. To implement the comparators, the simplest approach is to use two differential pairs followed by logic inverters, allowing a faster response as well as demanding less power consumption. In addition, a simple power reduction technique based on alternate comparator operation has been implemented.

To generate the bias current I_B, a β-multiplier approach has been developed which ensures a temperature and supply voltage independent bias current. This current can be next used in order to generate immunized comparison limits for the control circuit.

All the presented cells are suitable for a low-voltage low-power VFC, and therefore, they will be used in the design of complete VFCs in Chap. 4.

References

[AZC11]   Azcona, C., Calvo, B., Medrano, N., Bayo, A., Celma, S.: A 12-b enhanced input range on-chip quasi-digital converter with temperature compensation. IEEE Trans. Circuits Syst. II **58**(3), 164–168 (2011)
[BAK98]   Baker, R.J., Li, H.W., Boyee, D.E.: CMOS Circuit Design, Layout & Simulation. IEEE Press, New York (1997)
[BLA95]   Blalock, B.J., Allen, P.E.: A low-voltage, bulk-driven MOSFET current mirror for CMOS. In: Proceedings of the 1995 IEEE International Symposium on Technology, Circuits and Systems, (ISCAS '95), vol. 3, pp. 1972–1975. Seattle, Washington, 30 April to 3 May 1995 (1995)
[CAI95]   Cai, S.S.: Temperature-stable voltage-to-frequency converters. Master Thesis, University of Alberta, Canada (1995)
[CAL10]   Calvo, B., Medrano, N., Celma, S.: A full-scale CMOS voltage-to-frequency converter for WSN signal conditioning. In: Proceedings of IEEE International Symposium on Circuits and Systems (ISCAS'10), pp. 3088–3091. Paris, 30 May to 2 June 2010 (2010)
[CAR04]   de Carvalho-Ferreira, L.H., Pimenta, T.C.: An ultra low-voltage CMOS OTA Miller with rail-to-rail operation. In: Proceedings of the 16th International Conference on Microelectronics, (IMC'04), pp. 223–226. Taormina, Sicily, 25–27 Oct 2004 (2004)
[CHA10]   Chanapromma, C., Daoden, K.: A CMOS fully differential operational transconductance amplifier operating in sub-threshold region and its application. In: Proceedings of the 2nd International Conference on Signal Processing Systems (ICSPS'10), vol. 2, pp. 73–77. Dalian, 5–7 July 2010 (2010)
[CHE02]   Cheng, Y.: The influence and modeling of process variation and device mismatch for analog/RF circuit design. In: Proceedings of the Fourth IEEE International Caracas Conference on Devices, Circuits and Systems, vol. 46, pp. 1–8. Aruba, 17–19 April 2002 (2002)
[CHE05]   Chen, R.Y., Lin, S.F., Wu, M.S.: A linear CMOS voltage-to-current converter. In: Proceedings of the 2005 International Symposium on Signals, Circuits and Systems (ISSCS'05), vol. 2, pp. 677–680. Iasi, 14–15 July 2005 (2005)
[COM04]   Comer, D.J., Comer, D.T.: Using the weak inversion region to optimize input stage design of CMOS Op Amps. IEEE Trans. Circuits Syst. II: Express Briefs **51**(1), 8–14 (2004)
[COR03]   McCorquodale, M.S., Ding, M.K., Brown, R.B.: A CMOS voltage-to-frequency linearizing preprocessor for parallel plate RF MEMs varactors. In: Proceedings of the 2003 IEEE Radio Frequency Integrated Circuits (RFIC) Symposium, vol. 1, pp. 535–538. Philadelphia, PA, 8–10 June 2003 (2003)
[COR06]   Corbishley, P., Rodriguez-Villegas, E.: A low power low voltage rectifier circuit. In: Proceedings of the 49th IEEE International Midwest Symposium on Circuits and Systems (MWCAS'06), pp. 512–515. San Juan, 6–9 Aug 2006 (2006)
[CRA92]   Crawley, P.J., Roberts, G.W.: High-swing MOS current mirror with arbitrarily high output resistance. Electron. Lett. **28**(4), 361–363 (1992)
[DOU04]   Douglas, E.L., Lovely, D.F., Luke, D.M.: A low-voltage current-mode instrumentation amplifier designed in a 0.18-micron CMOS technology. In: Proceedings of the 2004

Canadian Conference on Electrical and Computer Engineering, vol. 3, pp. 1777–1780. Ontario, 2–5 May 2004 (2004)

[ENZ96]   Enz, C.C., Vittoz, E.A.: CMOS low-power analog circuit design. In: Cavin R.K., Liu W. (eds) Emerging Technologies: Designing Low Power Digital Systems, pp. 79–133. IEEE Press, New York (1996)

[GRA10]   Gray, P.R., Hurst, P.J., Lewis, S.H., Meyer, R.G.: Analysis and Design of Analog Integrated Circuits. Wiley, Asia (2010)

[GRE07]   Gregoire, B.R., Moon, U.K.: Process-independent resistor temperature-coefficients using series/parallel and parallel/series composite resistors. In: Proceedings of the 2007 IEEE International Symposium on Circuits and Systems (ISCAS'07), pp. 2826–2829. New Orleans, 27–30 May 2007 (2007)

[HAS11]   Hassen, N., Gabbouj, H.B., Besbes, K.: Low-voltage, high-performance current mirrors: Application to linear voltage-to-current converters. Int. J. Circuit Theory Appl. **39**(1), 47–60 (2011)

[HUN99]   Hung, C.C., Ismail, M., Halonen, K., Porra, V.: A low-voltage rail-to-rail CMOS V-I converter. IEEE Trans. Circuits Syst. II **46**(6), 816–820 (1999)

[JOH97]   Johns, D., Martin, K.W.: Analog Integrated Circuit Design. Wiley, New York (1997)

[LIM04]   Lim, Q.S.I., Kordesch, A.V., Keating, R.A.: Performance comparison of MIM capacitors and metal finger capacitors for analog and RF applications. In: Proceedings of the RF and Microwave Conference (RFM 2004), pp. 85–89. Subang Jaya, 5–6 Oct 2004 (2004)

[LOP07]   López-Martín, A.J., Ramírez-Angulo, J., Carvajal, R.G.: ±1.5 V 3 mW CMOS V-I converter with 75 dB SFDR for 6 V_{pp} input swings. Electron. Lett. **43**(6), 31–32 (2007)

[MAL01]   Maloberti, F.: Analog Design for CMOS VLSI Systems. Kluwer Academic, Boston (2001)

[RAM04]   Ramírez-Angulo, J., Carvajal, R.G., Torralba, A.: Low supply voltage high-performance CMOS current mirror with low input and output voltage requirements. IEEE Trans. Circuits Syst. II: Express Briefs, **51**(3), 124–129 (2004)

[RAM94]   Ramírez-Angulo, J.: Current mirrors with low input and low output voltage requirements. In: Proceedings of the 37th Midwest Symposium on Circuits and Systems (MWSCAS'94), vol. 1, pp. 107–110. University of Southern Louisiana, Layfayette, LA, 3–5 Aug 1994 (1994)

[ROD02]   Rodriguez-Villegas, E., Payne, A.J., Toumazou, C.: A 290 nW, weak inversion, Gm-C biquad. In: Proceedings of the IEEE International Symposium on Circuits and Systems (ISCAS'02), vol. 2, pp. 221–224. Scottsdale, AZ, 26–29 May 2002 (2002)

[SHU04]   Shukla, R., Ramírez-Angulo, J., López-Martín, A., Carvajal, R.G.: A low voltage rail to rail V-I conversion scheme for applications in current mode A/D converters. In: Proceedings of the 2004 International Symposium on Circuits and Systems (ISCAS'04), vol. 1, pp. 916–919. Vancouver, BC, 23–26 May 2004 (2004)

[SRI05]   Srinivasan, V., Chawla, R., Haster, P.: Linear current to voltage and voltage to current converters. In: Proceedings of the 48th Midwest Symposium on Circuits and Systems, (MWSCAS'05), pp. 675–678. Radisson Hotel, Cincinnati, OH, 7–10 Aug 2005 (2005)

[STE08]   Stefanović, D., Kayal, M.: Structured Analog CMOS Design. Springer, London (2008)

[UMC12]   United Microelectronics Corporation web page. http://www.umc.com/English/ (2012)

[VER95]   Vervoort, P.P., Wassenarr, R.F.: A CMOS rail-to-rail linear VI-converter. In: Proceedings of the 1995 IEEE International Symposium on Circuits and Systems (ISCAS'95), pp. 825–828. Seattle, Washington, 30 April to 3 May 1995 (1995)

[VIE07]   Vieira, F.C.B., Prior, C.A., Rodrigues, C.R., Perin, L., Martins, J.B.S.: Current mode instrumentation amplifier with rail-to-rail input and output. In: Proceedings of the 20th Annual Conference on Integrated Circuits and Systems Design, pp. 48–52. Rio de Janeiro, 3–6 Sept 2007 (2007)

[VIE08]   Vieira, F.C.B., Prior, C.A., Rodrigues, C.R., Perin, L., Martins, J.B.D.S.: Current mode instrumentation amplifier with rail-to-rail input and output. Analog Integr. Circuits Signal Process. **57**, 29–37 (2008)

[VIT09]   Vittoz, E.A.: Weak inversion for ultra low-power and very low-voltage circuits. In: Proceedings of the 2009 IEEE Asian Solid-State Circuits Conference (A-SSCC'09), pp. 129–132. Taipei, 16–18 Nov 2009 (2009)

[WAN06a]   Wang, C.C., Lee, T.J., Li, C.C., Hu, R.: An all-MOS high-linearity voltage-to-frequency converter chip with 520-kHz/V sensitivity. IEEE Trans. Circuits Syst. II **53**(8), 744–747 (2006)

[WAN06b]   Wang, A., Calhoun, B.H., Chandrakasan, A.P. Sub-Threshold Design for Ultra Low-Power Systems. Springer, New York (2006)

[YOU97]   You, F., Embabi, H.K., Duque-Carrillo, J.F., Sánchez-Sinencio, E.: An improved tail current source for low voltage applications. IEEE J. Solid-State Circuits **32**(8), 1173–1180 (1997)

Chapter 4
VFC Architectures

Chapter 4 presents four different VFC circuits based on the basic building blocks carried out in the previous chapter, complying with the specifications required for its use within a cost-effective microcontroller-based multi-sensor measurement system. Those include full-range input, low power consumption, supply voltages compatible with batteries, moderate linearity, temperature and supply insensitivity, and output frequency range suitable to be read by the embedded microcontroller.

All have been designed in the 0.18-μm CMOS process from UMC. The first configuration, VFC1, is supplied at the nominal voltage of the technology, $V_{DD} = 1.8$ V. It is based on an enhanced V-I converter and presents programmability, temperature compensation and a sleep-mode operation enable terminal. The other VFCs are supplied below the nominal value, at $V_{DD} = 1.2$ V, to prove their feasibility in downscaled technologies. All of them are based on the rail-to-rail feedback voltage attenuation (FBVA) V-I converter, and take account of temperature compensation. VFC2 is a single-input scheme, which includes an offset current. VFC3 is differential, whereas VFC4 can be configured in single/differential operation; both are compensated against supply voltage variations.

At the end of the chapter conclusions are drawn and a comparison with other architectures is done.

4.1 CMOS VFC

First, the proposed VFC architecture (VFC1) and the circuit design of the key function blocks are described. After that, main measured and simulated performances are summarized.

C.A. Murillo et al., *Voltage-to-Frequency Converters: CMOS Design and Implementation*,   81
Analog Circuits and Signal Processing, DOI 10.1007/978-1-4614-6237-8_4,
© Springer Science+Business Media New York 2013

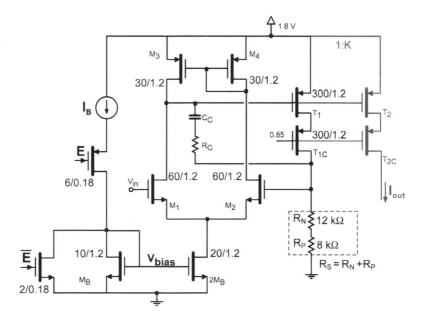

Fig. 4.1 Schematics of the V-I converter used in the VFC1

4.1.1   System Architecture

This first voltage-to-frequency converter consists of an enhanced V-I converter, followed by a conventional bidirectional current integrator driven by a voltage window comparator (VWC) as the control circuit. It is based in [AZC11], but including three different output frequency ranges. Programmability is a feature that adds versatility to the VFC allowing the possibility of being used with clocks that have other frequencies than the considered, 4 MHz. In addition, temperature compensation and the possibility to set the system into a sleep-mode have been included. The implementation of the main blocks which constitute the proposed VFC are described in the following.

V-I Converter. The complete V-I converter scheme, specifying transistor sizes and bias conditions is shown in Fig. 4.1. It is based on the subthreshold enhanced V-I converter analyzed in Sect. 3.1. Thus, the core circuitry consists of an OTA/ common-source amplifier driving a grounded resistor. The OTA is made up of a NMOS differential pair working in the subthreshold region. The bias current value, fixed to $I_B = 5$ µA, is introduced into the circuit through a simple current mirror M_B with 1:2 scaling factor. The resistor that makes the V-I conversion is set to $R_S = 20$ kΩ and it is implemented by the serial connection of $R_P = 8$ kΩ made up with a PND layer, and $R_N = 12$ kΩ made up with a HRP layer. The compensation network is a MIM capacitor $C_C = 0.4$ pF, and the parallel connection of an NMOS (2 µm/1.4 µm) and a PMOS (4 µm/1.4 µm) in triode region forming the compensation resistor R_C.

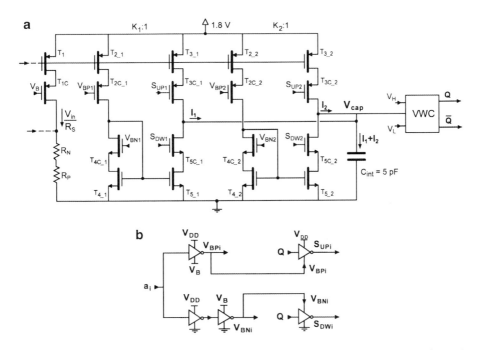

Fig. 4.2 (a) Programmable conventional bidirectional current integrator used in VFC1 and (b) its control circuit

Bidirectional Current Integrator. The complete bidirectional current integrator scheme is shown in Fig. 4.2a. It is the conventional current integrator analyzed in Sect. 3.2, but in this case it includes the programmable capability. Different output frequency ranges can be obtained easily by considering, instead of a single 1:K current scaling path, different digitally programmable 1:K_i current scaling mirrors. The digital control is made through a bit a_i which drives the gate voltage of the cascode transistors. Thus, the output current that contributes to voltage variation across the integrating capacitor C_{int} is proportional to the sum of all the currents corresponding to the active current mirrors, so that the time required to charge/discharge C_{int} between the limits V_H and V_L is given by

$$t = \frac{T}{2} = \sum \frac{C_{int}(V_H - V_L)}{a_i I_{in}/K_i} \tag{4.1}$$

Transistor sizes are specified in Table 4.1, where T_1-T_{1C} and R_N-R_P are those given in the V-I converter of Fig. 4.1. Two different current mirrors, denoted with

Table 4.1 Transistor sizes of the programmable bidirectional current integrator in VFC1

Transistor	W/L (μm/μm)
T_{2_1}–T_{2C_1}	11.25/1.2
T_{3_1}–T_{3C_1}	11.25/1.2
T_{4_1}–T_{4C_1}	3.75/1.2
T_{5_1}–T_{5C_1}	3.75/1.2
T_{2_2}–T_{2C_2}	22.5/1.2
T_{3_2}–T_{3C_2}	22.5/1.2
T_{4_2}–T_{4C_2}	7.5/1.2
T_{5_2}–T_{5C_2}	7.5/1.2

the subscript $_{i\,=\,1,\,2}$, have been considered. The bit a_i, as shown in Fig. 4.2b, activates or deactivates both, the scaled current (V_{BPi}, V_{BNi} driving T_{2_i}-T_{2C_i}, T_{4_i}-T_{4C_i}) and the bidirectional output current branches (S_{UPi}, S_{DWi} driving T_{3_i}-T_{3C_i}, T_{5_i}-T_{5C_i}).

Consider the case when $a_i = $ '0'. Then, the gate voltages of PMOS (T_{2C_i}) and NMOS (T_{4C_i}) cascode transistors in the current scaling path are respectively $V_{BPi} = V_{DD}$ and $V_{BNi} = 0$ V, so that both transistors are OFF. As a result, no scaled current is generated. In addition, the gate voltages of the PMOS (T_{3C_i}) and NMOS (T_{5C_i}) cascode transistors in the bidirectional output current branch are respectively $S_{UPi} = V_{DD}$ and $S_{DWi} = 0$ V independently of the VWC Q output logical level, keeping cut-off the output branch corresponding to a_i. On the other hand, when $a_i = $ '1', the gate voltages of PMOS (T_{2C_i}) and NMOS (T_{4C_i}) cascode transistors in the current scaling path are respectively $V_{BPi} = V_B$ and $V_{BNi} = V_B$. Consequently, the scaled current I_{in}/K_i is generated. Now, in the charging phase, as reported in Sect. 3.3, $Q = $ '1', so that $S_{UPi} = V_{BPi} \equiv V_B$ (ON) and $S_{DWi} \equiv$ '0' $\equiv 0$ V (OFF), while in the discharge phase $Q = $ '0', so that $S_{UPi} = $ '1' $\equiv V_{DD}$ (OFF) and $S_{DWi} = V_{BNi} \equiv V_B$ (ON), thus controlling the polarity of the bidirectional current integrator.

In this way, two different scaling paths are implemented, being the scaling factors $K_1 = 80/3$ and $K_2 = 40/3$. The integrating capacitor is implemented with a MIM capacitor of value $C_{int} = 5$ pF.

Control Circuit. The scheme of the control circuit comprising the output and the switching signals generation is shown in Fig. 4.3a. The complete scheme of the two comparators corresponding to the VWC specifying their transistor sizes is shown in Fig. 4.3b. Each comparator is a high-speed simple differential pair followed by inverter gates. The $\overline{RS}$ flip-flop is NAND based, with PMOS transistor sizes (3 μm/0.18 μm) and (1 μm/0.18 μm) for the NMOS.

The bias current value $I_B = 5$ μA is introduced into the circuit through a simple current mirror M_B with 1:4 scaling factor. Comparison limits are set to $V_H = 1.2$ V and $V_L = 0.6$ V and they are generated externally. In this way, the capacitor voltage V_{cap} swings from 0.6 to 1.2 V, keeping transistors in the bidirectional current integrator properly working in saturation. An enable terminal driven by Q controls two switches M_{SH} and M_{SL}, so that in the charging phase only the high comparator is active while in the discharging phase only the low one works.

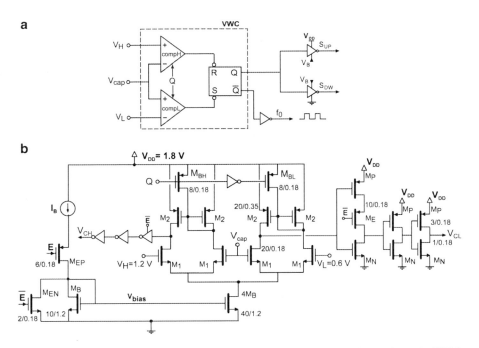

Fig. 4.3 Schematics of: (**a**) control circuit and (**b**) the two comparators that made up the VWC in the VFC1

Bias Circuit. The bias current I_B has been generated by means of a PMOS (1.5 μm/ 2 μm) whose gate is driven by an external voltage. This current is then replicated with different scaling factors to bias the V-I converter and the control circuit, as it has been explained in their respective sections. Gate voltages V_{BN} and V_{BP} that drives the cascode transistors gate are chosen to be the same for simplicity, and they are fixed in 0.8 V. This voltage is generated with a PMOS transistor (2 μm/5 μm) in diode connection. To monitor the current value, a replica bias circuit has been integrated. Thus, the bias current is fixed measuring the current in the replica circuit.

Enable Terminal. To set the whole system into a sleep-mode, an enable E has been included in both, the control circuit and the bias circuit. In the VWC, when the enable is active ($E = $ '1') the VWC is in ON; when the enable is deactivated ($E = $ '0',) the enable creates an open circuit in the first inverter (see Fig. 4.3), being transistors M_E in OFF, fixing the digital state of the remaining inverters. In the bias circuit (see Figs. 4.1 and 4.3) when the enable is active ($E = $ '1'), M_{EN} is in OFF and M_{EP} is in ON, so the bias current is generated and replicated to feed the V-I converter and the control circuit. When the enable is deactivated ($E = $ '0'), M_{EN} is in ON, making that V_{bias} is connected to ground, setting bias transistors M_B in OFF, whereas M_{EP} is in OFF, so no current is generated due to the open circuit; therefore, no current feeds the system. Thus, thanks to the enable terminal, the power consumption is drastically reduced.

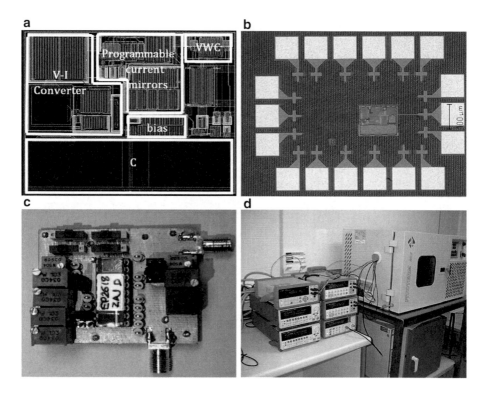

Fig. 4.4 (**a**) Layout, (**b**) microphotograph and (**c**) PCB of the VFC1, and (**d**) thermal chamber experimental set-up

Output Frequency. The output frequency expression of this programmable VFC is given by

$$f_0 = \frac{1}{2C_{int}R_S(V_H - V_L)} \sum_i a_i \frac{V_{in}}{K_i} \qquad (4.2)$$

being $i = 1$, 2. Ideal values of the parameters are $C_{int} = 5$ pF, $R_S = 20$ kΩ, $V_H = 1.2$ V, $V_L = 0.6$ V, $K_1 = 80/3$ and $K_2 = 40/3$. Thereby, the theoretical sensitivities are 312.5 kHz/V, 625 kHz and 937.5 kHz, for the three different output ranges $(a_1, a_2) = (1,0)$, $(a_1, a_2) = (0,1)$ and $(a_1, a_2) = (1,1)$. So, for the expected input operating range of [0.1, 1.6 V] the output frequency ranges would be [31.25, 500 kHz], [62.5 kHz, 1 MHz] and [93.75 kHz, 1.5 MHz].

4.1.2 Performances

The layout, the chip photograph and the PCB picture are shown in Fig. 4.4. For the experimental test, an inverter is added as a buffer at the VWC output to not modify the control signal when measuring f_0 (Fig. 4.3a), and a source follower buffer is

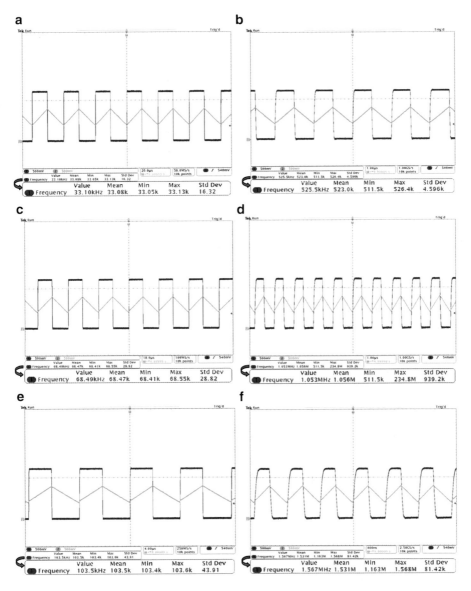

Fig. 4.5 Output signal and capacitor voltage for: (**a**) $V_{in} = 0.1$ V and (**b**) $V_{in} = 1.6$ V being $(a_1,a_2) = (1,0)$; (**c**) $V_{in} = 0.1$ V and (**d**) $V_{in} = 1.6$ V being $(a_1,a_2) = (0,1)$; (**e**) $V_{in} = 0.1$ V and (**f**) $V_{in} = 1.6$ V being $(a_1,a_2) = (1,1)$

connected to extract the signal V_{cap} across capacitor. In addition, an initial capacitor charge circuit has been included, that sets $V_{cap,0} = V_L$. With $E = $ '1', the circuit consumes less than 0.4 mW for a single supply of $V_{DD} = 1.8$ V. In sleep-mode, with $E = $ '0', the VFC has a power consumption of only 35 nW.

Figure 4.5 shows the output of the VFC and the capacitor voltage V_{cap} in the time domain, measured with a DPO 4104 oscilloscope from Tektronix, for input voltages $V_{in} = 0.1$ V and $V_{in} = 1.6$ V, with $(a_1,a_2) = (1,0)$, $(a_1,a_2) = (0,1)$ and $(a_1,a_2) = (1,1)$. It can be seen that V_{cap} varies correctly between V_H and V_L. The corresponding output frequencies are 33.10 kHz and 525.5 kHz, 68.49 kHz and 1.053 MHz, and 103.5 kHz and 1.567 MHz, respectively. The distortion in the rise and fall times appears as the frequency increases due to output parasitic capacitances.

Figure 4.6 shows the output frequency vs. the input voltage for the three output frequency ranges measured with the pulse counter Agilent 53132A. To prove the VFC feasibility for the target application, it is also shown the output frequency vs. the input voltage for the three output ranges with their relative errors, obtained by introducing the signal into a digital port of the ATMega 1281 µC −suitable for WSN nodes−, and performing the digitalization using the direct counting method (DCM) with $f_{clk} = 4$ MHz and $T_W = 16.4$ ms In both cases, for a [0.1, 1.6 V] input range, the output range varies linearly between the corresponding limits and the differences are negligible. The linearity error is kept below 0.029 % and a maximum relative error of 4.2 % is obtained over all the sensitivity cases.

To make measurements against temperature variations, the prototype has been introduced into a Fitorterm 22E thermal chamber from Aralab.

Figure 4.7 shows the output frequencies and the maximum relative error over the three input ranges, varying the temperature from −40 to +120 °C. Sensitivities remain almost constant (± 1.7 % in the worst case: 106 ppm/°C), being the maximum relative error of 6.6 %. However, the linearity error is increased to 0.047 % in the worst case $(a_1,a_2) = (1,0)$ at +120 °C. This is probably due to transistor T_1 entering in the triode region for $V_{in} = 1.6$ V at high temperature, what causes an increase in the linearity error.

The main performances of the VFC1 are summarized in Table 4.2.

4.2   CMOS Rail-to-Rail VFC

The previously proposed VFC covers almost, but not completely, the 0 to V_{DD} input range. Therefore, a rail-to-rail VFC will be presented in this section. In addition, despite the power reduction achieved with respect to the existing VFCs, in this second design we will take further advantage of the subthreshold operation, designing an ultra-low-power VFC, which will be denominated VFC2.

First, the proposed VFC2 architecture and the circuit design of the key function blocks will be described. After that, main measured performances are summarized.

4.2.1   System Architecture

The VFC2 consists of a rail-to-rail FBVA V-I converter, followed by a low-power bidirectional current integrator driven by a VWC. Its distinctive features, besides rail-to-rail operation, are temperature compensation and the addition of an offset

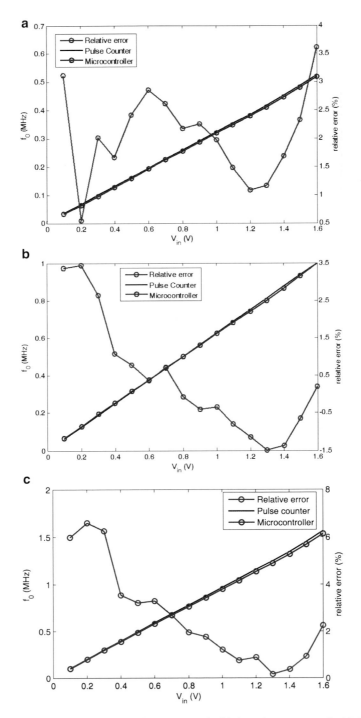

Fig. 4.6 Output frequency vs. input voltage measured with the pulse counter and with the microcontroller and its relative error for: (**a**) $(a_1,a_2) = (1,0)$, (**b**) $(a_1,a_2) = (0,1)$ and (**c**) $(a_1,a_2) = (1,1)$

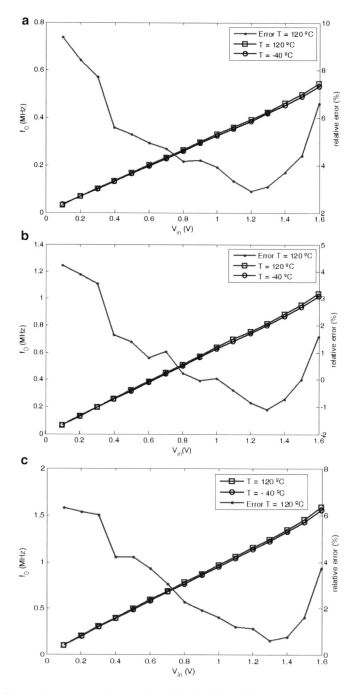

Fig. 4.7 Output frequency vs. input voltage for the three different frequency ranges and for temperatures of −40 and +120 °C

Table 4.2 Summary of the main VFC1 performances

Parameter	$(a_1, a_2) = (1, 0)$	$(a_1, a_2) = (0, 1)$	$(a_1, a_2) = (1, 1)$
Technology	0.18 μm CMOS	0.18 μm CMOS	0.18 μm CMOS
Supply voltage (V)	1.8	1.8	1.8
Input range (V)	0.1–1.6	0.1–1.6	0.1–1.6
Output range	32.2–517.7 kHz	65.1–1001 kHz	0.097–1.534 MHz
Linearity error (%)	0.029	0.016	0.022
Linearity error $(-40, +120\ {}^{\circ}C)$ (%)	0.039	0.032	0.047
Sensitivity (kHz/V)	319.6	616.6	939.9
Relative error $(-40, +120\ {}^{\circ}C)$ (%)	3.6	3.4	6.6
Power consumption (mW)	0.36	0.36	0.36

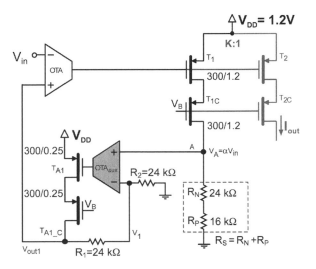

Fig. 4.8 Schematics of the V-I converter used in the VFC2

output frequency. The offset frequency is generated by introducing an offset current into the bidirectional current integrator.

V-I Converter. The complete V-I converter scheme, specifying transistor sizes is shown in Fig. 4.8. It is based on the rail-to-rail FBVA V-I converter analyzed in Sect. 3.1, so for detailed information refer to this section.

The core circuitry consists of an OTA/common-source amplifier with a floating dynamic battery, formed by an auxiliary OTA/common-source amplifier and feedback resistors $R_1 = R_2$ in its negative feedback loop, thus being the attenuation factor $\alpha = 0.5$. The main OTA has two input stages in parallel whereas the auxiliary OTA has only a PMOS input differential pair. Both are shown in Fig. 4.9, including transistor (W/L) sizes in (μm/μm). Transistors in both OTAs work in the subthreshold region.

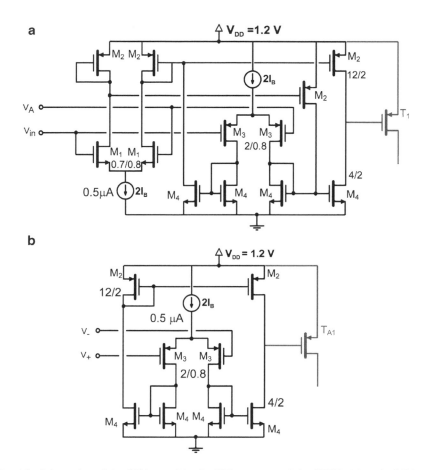

Fig. 4.9 Schematics of the OTAs used in the V-I converter of the VFC2: (**a**) main OTA and (**b**) OTA_{aux}

The bias current, fixed to 0.5 μA, is introduced into the circuit through current mirrors with 1:1 scaling factor, being (10 μm/1 μm) their dimensions. To minimize the circuit temperature dependence, the resistor that makes the main V-I conversion is set to $R_S = 40$ kΩ, and it is implemented by the serial connection of $R_P = 16$ kΩ (PND layer) and $R_N = 24$ kΩ (HRP layer). HRP resistors in the dynamic battery are set to $R_1 = R_2 = 24$ kΩ. The compensation networks are made up with MIM capacitors, $C_C = 0.88$ pF and $C_{C\_A} = 20$ fF; and R_C, R_{C_A} are the parallel connection of NMOS (2 μm/1.4 μm) and PMOS (4 μm/1.4 μm) transistors.

Bidirectional Current Integrator. The complete bidirectional current integrator scheme is shown in Fig. 4.10 specifying its transistor sizes, except T_1-T_{1C} and R_N-R_P that are given together in the V-I converter (Fig. 4.8). It is the low-power current integrator analyzed in Sect. 3.2, but including an offset current I_0 that is the responsible for the output frequency offset. The scaling factor K between transistors

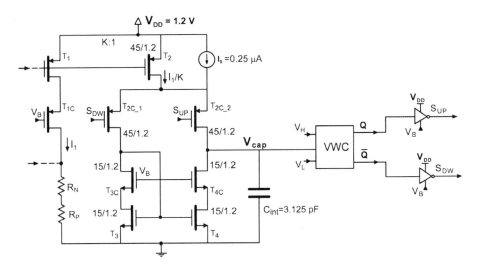

Fig. 4.10 Low-power bidirectional current integrator used in VFC2

T_1 and T_2 is $K = 20/3$, the MIM integrating capacitor is set to $C_{int} = 3.125$ pF and the offset current to $I_0 = 0.25$ µA (ideal values).

As an offset current is introduced into the circuit, the current that contributes to voltage variation across capacitor is proportional to the sum of the scaled current plus the offset current. In this way, the time required to charge/discharge the capacitor between the limits V_H and V_L is given by (4.3).

$$t = \frac{T}{2} = \sum \frac{C_{int}(V_H - V_L)}{(I_{in}/K + I_0)} \tag{4.3}$$

Control Circuit. The scheme of the control circuit is identical to that of VFC1: a VWC, made up of two alternatively working high-speed simple differential pair comparators followed by inverters and a NAND-based $\overline{RS}$ flip-flop. However, this circuit is supplied with a single voltage of 1.2 V, the bias current $I_B = 0.5$ µA is introduced into the circuit through a simple current mirror M_B with 1:5 scaling factor, comparison limits are set to $V_H = 0.8$ V and $V_L = 0.4$ V, and it does not include the enable terminal E. The complete scheme of the two comparators, specifying transistor sizes is shown in Fig. 4.11.

Bias Circuit and Comparison Limits Generation. Like in the VFC1, the bias current $I_B = 0.5$ µA has been generated by a PMOS (2 µm/5 µm) whose gate is driven by an external voltage. This current biases V-I, comparators, as well as serves to generate the cascode gate voltages $V_B = 0.4$ V. Comparison limits V_H and V_L are generated by a voltage divider made with PMOS transistors in diode connection (51 µm/3 µm).

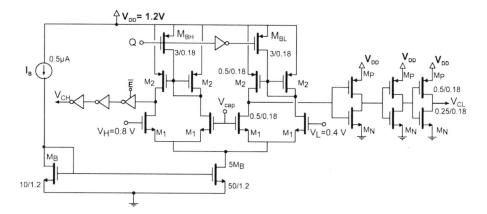

Fig. 4.11 Schematics of the two comparators that made up the VWC in the VFC2

Output Frequency. The output frequency expression of the VFC2 is given by

$$f_0 = \frac{1}{2C_{int}(V_H - V_L)}\left(\frac{\alpha V_{in}}{KR_S} + I_0\right) \tag{4.4}$$

where ideal values of the parameters have been given in the previous sections: $C_{int} = 3.125$ pF, $R_S = 40$ kΩ, $V_H = 0.8$ V, $V_L = 0.4$ V, $\alpha = 0.5$, $K = 20/3$ and $I_0 = 0.25$ µA. Thereby, the theoretical sensitivity is 750 kHz and the offset frequency is 100 kHz. So, for an input range of [0.0, 1.2 V] the output frequency range would be [0.1, 1.0 MHz].

4.2.2 Performances

The layout, including test buffers and start-up circuitry, and the microphotograph of the aforementioned proposed VFC are shown in Fig. 4.12. It consumes less than 70 µW. Figure 4.13 shows in the time domain the response of the VFC2 for input voltages $V_{in} = 0.0$, and 1.2 V. The corresponding output frequencies are 101.2, and 963.9 kHz, respectively. It can also be seen that the capacitor voltage, V_{cap}, varies correctly between $V_H = 0.8$ V and $V_L = 0.4$ V.

The VFC response all over the input voltage [0.0 − 1.2 V] has been measured both with an Agilent 53132A counter and with an ATMega 1281 µC using the DCM with $f_{clk} = 4$ MHz and $T_W = 16.4$ ms, obtaining similar results, shown in Fig. 4.14, together with the relative error. The corresponding output frequency range is [102.3, 964.1 kHz], being the experimental sensitivity 726.8 kHz/V (3.1 % error), the offset of 102.3 kHz (2.3 %), the linearity error 0.014 % (12 bits) and the maximum relative error of 3.2 %.

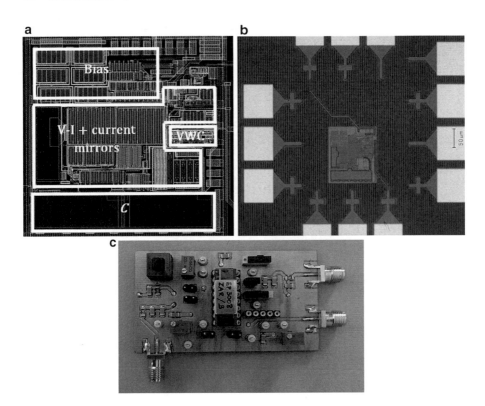

Fig. 4.12 (a) Layout, (b) microphotograph and (c) PCB of the VFC2

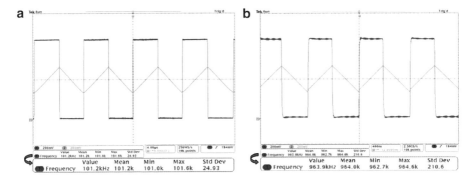

Fig. 4.13 Output signal and capacitor voltage for: (a) $V_{in} = 0.0$ V and (b) $V_{in} = 1.2$ V

The prototype has been validated against temperature variations, from -40 to $+120\,°C$, obtaining the results shown in Fig. 4.15. Over all this temperature range the sensitivity remains almost constant (± 1.5 % in the worst case: 94 pmm/$°C$), the worst relative error is 4.5 % and the maximum linearity error is 0.022 %.

The main performances of the VFC2 are summarized in Table 4.3.

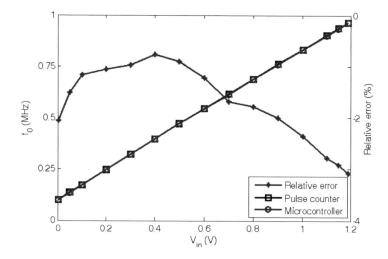

Fig. 4.14 Output frequency vs. input voltage measured with the pulse counter and with the microcontroller and its relative error

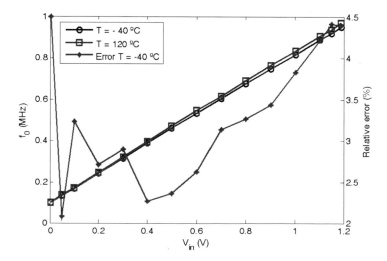

Fig. 4.15 Output frequency vs. input voltage for –40 and +120 °C, and the maximum relative error corresponding to $T = -40$ °C

Table 4.3 Summary of the main VFC2 performances

Parameter	Value
Technology	0.18 μm CMOS
Supply voltage	1.2 V
Input range	0.0–1.2 V
Output range	102.3–964.1 kHz
Linearity error	0.014 %
Linearity error (-40, $+120$ °C)	0.022 %
Sensitivity	726.8 kHz/V
Relative error (-40, $+120$ °C)	4.5 %
Power consumption	69 μW

4.3 Differential VFCs

There are several sensors with differential output. The most known ones are those where the signal is coming from a Wheastone bridge. Therefore, it is interesting to design a VFC which is capable to process signals in a differential way, which brings advantages in terms of better immunity to noise and interference. This issue is addressed in this section through the proposal of two differential VFCs, denominated VFC3 and VFC4.

4.3.1 VFC3: System Architecture

The VFC3 consists of a rail-to-rail differential V-I converter based on the rail-to-rail FBVA V-I converter followed by a conventional based bidirectional current integrator driven by a VWC control circuit. First, the circuit description and implementation of these blocks is carried out. After that, main simulated performances are summarized.

Differential V-I Converter. The complete differential V-I converter scheme is shown in Fig. 4.16. It consists of two rail-to-rail FBVA V-I converters, the FBVA$_1$ formed by OTA$_1$, OTA$_{aux1}$, T$_1$, T$_{1C}$, T$_{A1}$, T$_{A1C}$, R_1 and R_2, and the FBVA$_2$ formed by OTA$_2$, OTA$_{aux2}$, T$_3$, T$_{3C}$, T$_{A2}$, T$_{A2C}$, R_3 and R_4. Transistor sizes are shown in Table 4.4.

The FFVA$_1$ transforms the input signal V_{in+} into a signal V_A at node A, given by

$$V_A = \frac{R_2 V_1 + R_1 V_{in+}}{R_1 + R_2} \tag{4.5}$$

The voltage level V_1 is fixed to the common mode voltage V_{CM} and resistors in the feedback loop are set to $R_1 = R_2$, which sets $\alpha = 0.5$. Taking into account that $V_{in+} = V_{CM} + V_d/2$, the voltage at node A remains as following:

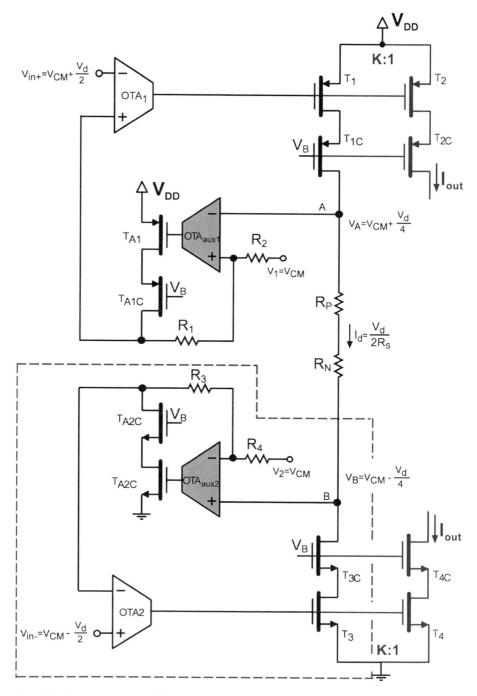

Fig. 4.16 Schematics of the V-I converter used in the VFC3

Table 4.4 Transistor sizes of the differential V-I converter of VFC3

Transistor	W/L (μm/μm)
T_1–T_{1c}	300/1.2
T_3–T_{3c}	300/1.2
T_{A1}–T_{A1C}	50/0.25
T_{A2}–T_{A2C}	50/0.25

$$V_A = 2\alpha V_{CM} + \alpha \frac{V_d}{2} = V_{CM} + \frac{V_d}{4} \qquad (4.6)$$

Similarly, FFVA$_2$ transforms the input signal V_{in-} into a signal V_B at node B, given by

$$V_B = \frac{R_4 V_2 + R_3 V_{in-}}{R_3 + R_4} = V_{CM} - \frac{V_d}{4} \qquad (4.7)$$

again selecting the auxiliary voltage level $V_2 = V_{CM}$ and feedback resistors $R_3 = R_4$.

This results in a fully symmetric structure, which maintains at nodes A and B the common mode voltage V_{CM}, while the differential voltage V_d is halved. Therefore, the voltage across the temperature compensated resistor R_S ($R_S = R_N + R_P$) is $V_A - V_B = V_d/2$, and the generated current is $I_d = V_d/2R_S$.

As the input of both OTAs in the FBVA$_1$ swing between V_{CM} and V_{DD} they could be made by using a NMOS input stage OTA. In the same way, as the input of both OTAs in FBVA$_2$ swing between 0 V and V_{CM} they could be made by using a PMOS input stage OTA. However, to have a fully symmetrical and modular design, with less mismatching, rail-to-rail OTAs are used for both OTA and OTA$_{aux}$ in FBVA$_1$ and FBVA$_2$. This OTA is the same used as in VFC2, and its scheme and dimensions are shown in Fig. 4.9a.

The common mode voltage is selected to be $V_{CM} = 0.6$ V. The bias currents are fixed to 0.5 μA and they are introduced into the circuit with simple current mirrors of dimensions (10 μm/1 μm) with 1:1 scaling factor. The V-I conversion resistor is $R_S = 40$ kΩ ($R_{PND} = 16$ kΩ and $R_{HRP} = 24$ kΩ). Resistors in the feedback loops are set to $R_1 = R_2 = R_3 = R_4 = 24$ kΩ and they are implemented by means of HRP layers. In both, FBVA$_1$ and FBVA$_2$, the compensation network is set to $C_{CI} = C_{C2} = 2$ pF, $C_{CI\_A} = C_{C2\_A} = 200$ fF and the compensation resistors are made by a NMOS (2 μm/1.4 μm) and a PMOS (4 μm/1.4 μm) in parallel.

Bidirectional Current Integrator. The complete bidirectional current integrator scheme is shown in Fig. 4.17 specifying its transistor sizes. The current signal $I_d = V_d/2R_S$ is directly replicated with a scaling factor K:1 being $K = 20/3$ through T_2-T_{2C} and T_4-T_{4C}, with T_{3C} and T_{4C} acting as switches. In the charging phase, $Q = $ '1', $S_{UP} = V_B$ and $S_{DW} = 0$ V, thus having active the branch T_2-T_{2C}, whereas in the discharging phase, $Q = $ '0', $S_{UP} = V_{DD}$ and $S_{DW} = V_B$, thus having active the branch T_4-T_{4C} to drive the MIM capacitor $C_{int} = 3.125$ pF (ideal value).

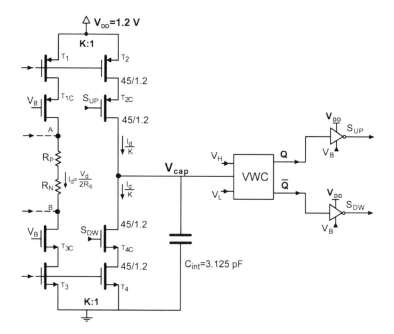

Fig. 4.17 Bidirectional current integrator used in VFC3

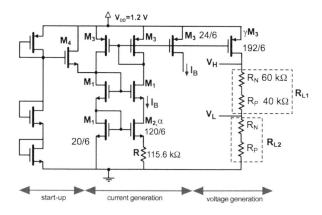

Fig. 4.18 Schematics of the bias currents generation for the VFC3

Control Circuit. The control circuit is identical to the one used in VFC2, and shown in Fig. 4.11. Comparison limits, set to $V_H = 0.8$ V and $V_L = 0.4$ V, are internally generated as explained in the next section.

Bias Circuit. The bias circuit scheme is shown in Fig. 4.18. It is a β-multiplier working in the subthreshold region, which generates the supply independent bias current $I_B = 500$ nA according to following expression

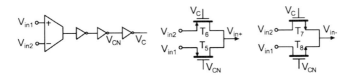

Fig. 4.19 Sign circuit

$$I_B = \frac{nV_T}{R} \ln(\alpha) \tag{4.8}$$

extracted in Chap. 3, Sect. 3.4, with $\alpha = 6$, and the resistor R is set to 115.6 kΩ. As shown also in that section, R is made with the serial connection of a NWELL and a PND resistors, being $R_{PND} = 52$ kΩ and $R_{NWELL} = 63.6$ kΩ, to achieve a temperature- and supply-independent bias current. This bias current is then replicated with different scaling factors to generate the current needed for the V-I converter, the control circuit, the common mode voltage generation V_{CM} and $V_B = 0.6$ V. In addition, comparison limits V_H and V_L are generated using the current I_B replicated with a scaling factor 1:8, and temperature-independent resistors $R_{L1} = R_{L2} = 100$ kΩ ($R_{PND} = 40$ kΩ and $R_{HRP} = 60$ kΩ) being the comparison limits $V_L = 8I_B R_L = 0.4$ V and $V_H = 8I_B 2R_L = 0.8$ V.

The deviation of the bias current I_B over a $1.0 - 1.4$ V variation in the supply voltage is 4.5 nA/V, and over a $(-40, +120\,^\circ\text{C})$ temperature range is 52.5 pA/$^\circ$C. As for V_H–V_L, it shows a variation of 11.3 mV/V over a $1.0 - 1.4$ V supply voltage variation and a variation of 29.4 μV/$^\circ$C over the temperature range $(-40, +120\,^\circ\text{C})$.

Sing Circuit. To assure the proper direction of the current through R_S, a sign circuit can be required. It is implemented by means of a rail-to-rail comparator and transistors acting as switches, as shown in Fig. 4.19. The comparator is made up with an open-loop OTA as those used in the V-I converter, followed by inverters. When $V_{in1} > V_{in2}$, $V_C = V_{DD} \equiv$ '1' and $V_{CN} = 0$ V $\equiv$ '0', so that T_5 and T_7 are on and T_6 and T_8 are off; therefore, $V_{in+} = V_{in1}$, $V_{in-} = V_{in2}$. Conversely, when $V_{in1} < V_{in2}$, $V_{in+} = V_{in2}$ and $V_{in-} = V_{in1}$.

Output Frequency. The output frequency expression of this dVFC is given by

$$f_0 = \frac{1}{2C_{int}(V_H - V_L)} \left(\frac{V_{in+} - V_{in-}}{2KR_S} \right) \tag{4.9}$$

where ideal values of the parameters have been given in the previous section: $C_{int} = 3.125$ pF, $V_H = 0.8$ V, $V_L = 0.4$ V, $K = 20/3$ and $R_S = 40$ kΩ. Thereby, the theoretical sensitivity is 750 kHz, so for a differential input range of $[0.6 \pm 0.6$ V$]$ the output frequency range would be $(0.0, 0.9$ MHz$)$.

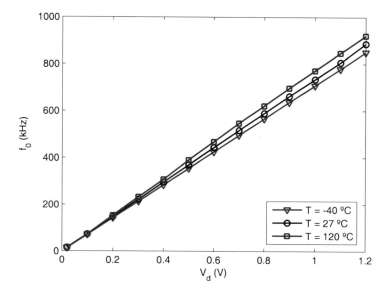

Fig. 4.20 Output frequency vs. input voltage for −40 and +120 °C

4.3.2   VFC3: Performances

This VFC has been simulated using SPECTRE with a BSIM3v3 level 53 transistor model. The overall circuit consumes less than 65 μW.

At room temperature for a differential input range of [0.6 ± 0.6 V], the output range varies linearly between 0.0 and 900 kHz, with an experimental sensitivity of 736.5 kHz/V (1.8 % error), a linearity error of 0.002 % and a maximum relative error of 2.8 %.

Simulations against temperature variations over a (−40, +120 °C) are shown Fig. 4.20. Over that range, the maximum relative error is 6.7 %, the maximum sensitivity error is 5.6 % (350 ppm/°C) and the linearity error remains below 0.013 %.

When the system is simulated for 30 % supply voltage variations (1.2 ± 0.2 V) being $V_{CM} = V_{DD}/2$ V, as shown in Fig. 4.21 the input range varies accordingly. However, the errors remain bounded: the maximum relative error is 10 %, the maximum gain error is 4.6 % and the linearity error remains below 0.010 %.

The VFC3 has also been tested for $V_{CM} = 0.6 \pm 0.3$ V variations at the nominal $V_{DD} = 1.2$ V supply voltage, remaining the frequency nearly constant with a maximum variation of 0.4 % with respect to the frequency at $V_{CM} = 0.6$ V.

The main performances of the VFC2 are summarized in Table 4.5.

4.3.3   VFC4: System Architecture

The VFC4 consists of a rail-to-rail differential V-I converter based on the rail-to-rail FBVA V-I converter followed by a low-power bidirectional current integrator

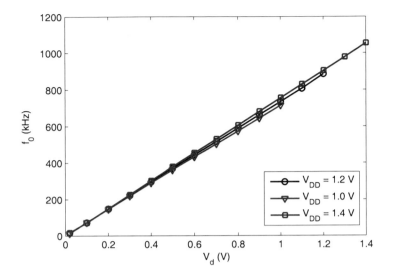

Fig. 4.21 Output frequency vs. input voltage for a supply voltage variation of $(1.0 - 1.4 \text{ V})$

Table 4.5 Summary of the main VFC3 performances

Parameter	Value
Technology	0.18 μm CMOS
Supply voltage	1.2 V
Input range	0.0–1.2 V
Output range	0.0–0.9 MHz
Linearity (room temperature)	0.002 %
Sensitivity (room temperature)	736.5 kHz/V
Sensitivity error (-40, $+120$ °C)	5.6 %
Linearity error (-40, $+120$ °C)	0.010 %
Power consumption	65 μW

driven by a control circuit. It has two operation modes: single input and differential input. First, the implementation of these blocks is described. Then, main simulated performances are summarized.

Differential V-I converter. Figure 4.22 shows the complete differential V-I converter scheme in a complete VFC scheme, whereas Table 4.6 offers transistor sizes. This circuit consists of two rail-to-rail FBVA V-I converters, the FBVA$_1$ formed by OTA$_1$, OTA$_{\text{aux1}}$, T$_1$, T$_{1C}$, T$_{A1}$, T$_{A1C}$, R_1 and R_2, and the FBVA$_2$ formed by OTA$_2$, OTA$_{\text{aux2}}$, T$_5$, T$_{5C}$, T$_{A2}$, T$_{A2C}$, R_3 and R_4.

FBVA$_1$ attenuates the input voltage at node A, being V_A given by

$$V_A = \frac{R_1 V_{in+}}{R_1 + R_2} = \alpha V_{in+} \tag{4.10}$$

Resistors in the feedback loop are set to $R_1 = R_2$. This means that $\alpha = 0.5$, and therefore $V_A = V_{in+}/2$. Similarly, for the FBVA$_2$, the voltage at node B is given by

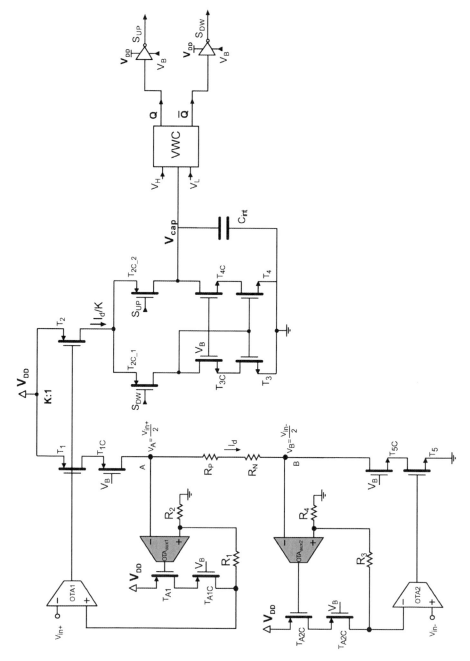

Fig. 4.22 Complete scheme of the VFC4

Table 4.6 Transistor sizes of the differential V-I converter used in VFC4

Transistor	W/L (μm/μm)
T_1-T_{1c}	300/1.2
T_3-T_{3c}	400/1.2
$T_{A1}-T_{A1C}$	400/0.25
$T_{A2}-T_{A2C}$	300/0.25

$$V_B = \frac{R_3 V_{in-}}{R_3 + R_4} = \alpha V_{in-} = \frac{V_{in-}}{2} \qquad (4.11)$$

When the VFC4 works in single mode, $V_{in-} = 0$, being the voltage across resistor R_S is $V_A - V_B = V_{in+}/2$, and thus, the generated current is $I_d = V_{in+}/2R_S$.

When the VFC4 works in differential input mode, $V_{in+} = V_{CM} + V_d/2$ and $V_{in-} = V_{CM} - V_d/2$. Therefore, the voltage across resistor R_S is $V_A - V_B = V_d/2$, and thus the generated current is $I_d = V_d/2R_S$.

In both cases, although the OTA design can be simplified according to the input swing of each active cell, because a dual operation is desired and to keep symmetry to minimize mismatch errors, all OTAs are implemented with the rail-to-rail OTA structure shown in Fig. 4.9a.

The bias currents, fixed to 0.5 μA, are introduced into the circuit with 1:1 current mirrors of dimensions (10 μm/1 μm). The resistor $R_S = 40$ kΩ ($R_{PND} = 16$ kΩ and $R_{HRP} = 24$ kΩ), and HRP feedback resistor $R_1 = R_2 = R_3 = R_4 = 24$ kΩ. Compensation networks are the same in both FFVAs: $C_{C1} = C_{C2} = 2$ pF, $C_{C1\_A} = C_{C2\_A} = 20$ fF and the compensation resistors are a NMOS (2 μm/1.4 μm) and a PMOS (4 μm/1.4 μm) in parallel. In differential mode, the common mode voltage is fixed to $V_{CM} = 0.6$ V.

Bidirectional Current Integrator. The low-power bidirectional current integrator used in VFC4 is the same than the one used in VFC2. It is reproduced in Fig. 4.9, with the only difference that this time the offset current is not included, as can be seen in Fig. 4.22.

Control Circuit, Bias Circuit and Sign Circuit. These circuits are the same than those used in VFC3 and are shown in Figs. 4.11, 4.18, and 4.19.

Output Frequency. The output frequency expression for VFC4 is given by

$$f_0 = \frac{1}{2C_{int}(V_H - V_L)}\left(\frac{V_{eff}}{2KR_S}\right) \qquad (4.12)$$

where $V_{eff} = V_{in+}$ in single mode, and $V_{eff} = V_{in+} - V_{in-}$ in differential mode. The ideal values of the parameters have been given in the previous sections: $C_{int} = 3.125$ pF, $V_H = 0.8$ V, $V_L = 0.4$ V, $K = 20/3$ and $R_S = 40$ kΩ. Thereby, the theoretical sensitivity is 750 kHz, so for a differential input range of $[0.6 \pm 0.6$ V], and for a single input range of $[0.0 - 1.2$ V], the output frequency ranges would be [0.0, 0.9 MHz].

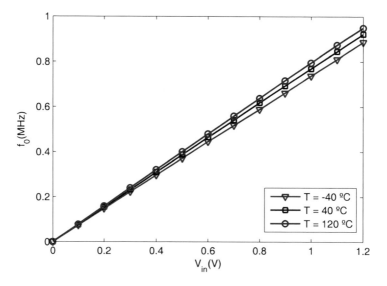

Fig. 4.23 Output frequency vs. input voltage for −40 and +120 °C

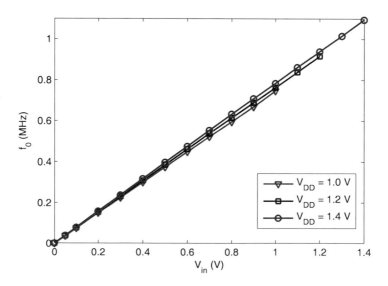

Fig. 4.24 Output frequency vs. input voltage for a supply voltage variation of (1.0 − 1.4 V)

4.3.4   *VFC4: Performances*

The VFC4, designed in UMC 0.18-μm CMOS technology feed with a single supply
of 1.2 V, has been simulated using SPECTRE with a BSIM3v3 level 53 transistor
model and its power consumption is below 75 μW.

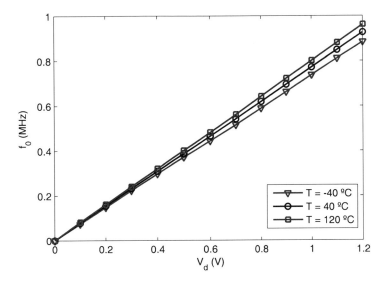

Fig. 4.25 Output frequency vs. input voltage for −40 and +120 °C

Single Mode. The VFC4 output frequency varies linearly from 0.0 to 0.9 MHz with a sensitivity of 762.7 kHz/V (1.7 % error), a linearity error of 0.007 % and a maximum relative error of 3.0 %.

Figure 4.23 shows the output frequency f_0 over the input range at different temperatures ranging from −40 to +120 °C, and Fig. 4.24 shows f_0 over the input range for different supply voltages.

Over all the (−40, +120 °C) temperature range, the maximum sensitivity error is 5.9 % (369 ppm/°C), the linearity error remains below 0.008 %, and the maximum relative error is 7.1 %, whereas against the 30 % of supply voltage variations (1.2 ± 0.2 V), the maximum sensitivity error is 4.1 %, the linearity error remains below 0.009 %, and the maximum relative error is 6.0 %.

Differential Mode. In differential mode, the sensitivity of VFC4 is 765.7 kHz/V (2.1 % error) with a linearity error of 0.001 %, and a maximum relative error of 3.1 %.

Over a temperature range from −40 to +120 °C, the VFC4 maximum sensitivity error is 6.6 % (412.5 ppm/°C), its linearity error is 0.004 % and the maximum relative error is 8.6 %. The obtained results are shown in Fig. 4.25.

Varying the supply voltage from 1.0 to 1.4 V, the maximum main errors are: 4.7 % sensitivity error, 0.002 % linearity error and 7.7 % relative error. Figure 4.26 shows f_0 for different supply voltages, setting $V_{CM} = V_{DD}/2$.

Against V_{CM} variations ($V_{CM} = 0.6 \pm 0.3$ V) the maximum variation in frequency is of 0.6 % with respect to the frequency at $V_{CM} = 0.6$ V.

The main performances of the VFC4 are summarized in Table 4.7.

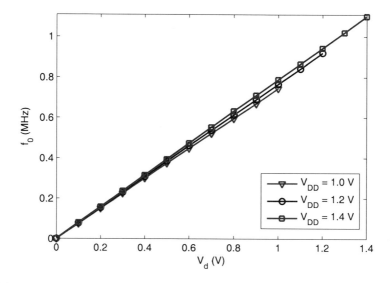

Fig. 4.26 Output frequency vs. input voltage for a supply voltage variation of (1.0 − 1.4 V)

Table 4.7 Summary of the main VFC4 performances

Parameter	Single mode	Differential mode
Technology	0.18 μm CMOS	0.18 μm CMOS
Supply voltage (V)	1.2	1.2
Input range (V)	0.0 – 1.2	0.6 ± 0.6
Output range (MHz)	0.0 – 0.9	0.0 – 0.9
Linearity (room temperature) (%)	0.007	0.001
Sensitivity (room temperature) (kHz/V)	762.7	765.7
Sensitivity error (−40, +120 °C) (%)	5.9	6.6
Linearity error (−40, +120 °C) (%)	0.009	0.004
Power consumption (μW)	75	75

4.4 Conclusions

In this chapter several voltage-to-frequency converters are proposed. The design of their basic blocks is presented and their main performances are summarized.

Section 4.1 presented a 1.8 V–0.18 μm CMOS programmable VFC. This VFC is very simple and compact, it exhibits an almost full input range and it is temperature compensated. Three frequency output ranges can be digitally selected, making this VFC suitable to be used with different clock frequencies. Its behavior against temperature variations has been tested in the temperature range of interest for the target application. In addition, there is the possibility to set the system into a sleep mode while no conversion is required.

Section 4.2 presented a 1.2 V–0.18 μm CMOS rail-to-rail VFC, which is temperature independent and includes an offset frequency. This ultra low-power VFC shows promising performances for its use in WSN environmental applications.

Section 4.3 proposed two 1.2 V–0.18 μm CMOS differential VFCs, suitable to accommodate differential signals. The first one, VFC3, works with differential signals while the second one, VFC4, can process both single and differential signals. They present highly linear response, keeping ultra-low power consumption, with a complete design that includes temperature and supply independent biasing.

Thus, different features have been included and validated in the presented VFCs: Single/differential mode operation, rail-to-rail input, temperature and supply voltage independence, programmability, offset, and a sleep-mode enable. All of these features can be included into a new VFC.

Next, two tables summarize the main performances of the presented VFCs, comparing them with other VFCs existing in the literature. Table 4.8 summarizes the single VFCs and Table 4.9 summarizes the differential VFCs.

Regarding the single operation VFCs, in addition to those collected in the table, there are others, mentioned in Chap. 1. The presented VFCs exhibit enhanced input range, coming to achieve rail-to-rail operation, while reducing drastically the power consumption and offering high linearity, and temperature immunity.

There are several dVFCs in the literature that do not appear in the table because they do not provide enough results or they give performances just applied to Wheatstone bridges, giving the data in terms of resistances. For instance, [JUS09, THE10a, THE10b] propose VFCs based on second-generation current conveyors (CCIIs), implemented with commercial devices, showing very limited differential input ranges: 0.2 V in the cases of [JUS09] and [THE10a], and 0.5 V in [THE10b]. [McD98] and [FER07] propose CMOS dVFCs. The dVFC in [McD98] has selectable sensitivities and the one in [FER07], despite the good performances in terms of gain error or linearity, has a power consumption of 85 mW. Thus, comparing with those that give the input range in terms of voltage, the proposed dVFCs are the only that achieve a rail-to-rail operation with lower power consumption and good linearity performances.

Table 4.8 Summary of integrated CMOS VFC circuits

	[WAN06]	[CAL09]	[CAL10]	[WAN07]	VFC1	VFC2	VFC4
Technology	0.25 μm CMOS	0.35 μm CMOS	0.35 μm CMOS	0.25 μm CMOS	0.18 μm CMOS	0.18 μm CMOS	0.18 μm CMOS
Supply (V)	2.5	3.0	3.0	2.5	1.8	1.2	1.2
Sensitivity	520 kHz/V	1.0 MHz/V	1.0 MHz/V	158.83 MHz/V	937.5 kHz/V	750 kHz/V	750 kHz/V
Input range (v)	0.1 – 0.8	1.0 – 2.0	0.1 – 2.7	0.0 – 0.9	0.1 – 1.6	0.0 – 1.2	0.0 – 1.2
Output frequency	52 – 416 kHz	1.2 – 2.2 MHz	0.1 – 2.7 MHz	0.0 – 55.4 MHz	0.097 – 1,500 kHz	0.1 – 1.0 MHz	0.0 – 0.9 MHz
Relative error (%)	<1.0	<0.7	<4.0	<8.5	6.6[a]	4.5[a]	7.1[a]
Linearity error (%)	–	<1	<4	–	0.047[a]	0.022[a]	0.008[a]
Power	–	1.03 mW	0.8 mW	>1 mW[b]	360 μW	69 μW	75 μW
Area (μm²)	517 × 596	–	440 × 460	–	137 × 100	140 × 150	–

[a]Over a (−40, +120 °C) T range
[b]Estimated power consumption

Table 4.9 Comparison of dVFC performances

Parameter	[PET10]	[VAL11]	VFC3	VFC4
Technology	Commercial devices	0.18 μm CMOS	0.18 μm CMOS	0.18 μm CMOS
Supply voltage (V)	±5	1.8	1.2	1.2
Sensitivity (kHz/V)	75	861	750	750
Input range	0.2 V diff (0.0 ± 0.1 V)	1.2 V diff (1.2 ± 0.6 V)	Full range (0.6 ± 0.6 V)	Full range (0.6 ± 0.6 V)
Relative error (%)	<5[a]	–	10[b]	8.6[b]
Linearity error (%)	–	0.4[a]	0.015[b]	0.004[b]
Power consumption (μW)	–	375	65	75

[a]Nominal
[b]For 30 % V_{DD} variation and (−40, +120 °C) T range

References

[AZC11]   Azcona, C., Calvo, B., Medrano, N., Bayo, A., Celma, S.: 12-b enhanced input range on-chip quasi-digital converter with temperature compensation. IEEE Trans. Circuits Syst. II Express Briefs **58**(3), 164–168 (2011)

[CAL09]   Calvo, B., Medrano, N., Celma, S., Sanz, M.T.: A low-power high-sensitivity CMOS voltage-to-frequency converter. In: Proceedings of the 52 IEEE International Midwest Symposium on Circuits and Systems (MWSCAS'09), pp. 118–121. Cancun, 2–5 Aug 2009 (2009)

[CAL10]   Calvo, B., Medrano, N., Celma, S.: A full-scale CMOS voltage-to-frequency converter for WSN signal conditioning. In: Proceedings IEEE International Symposium on Circuits and Systems (ISCAS'10), pp. 3088–3091. Paris, 30 May to 2 June 2010 (2010)

[FER07]   Ferrari, V., Ghisla, A., Vajna, Zs, K., Marioli, D., Taroni, A.: ASIC front-end interface with frequency and duty cycle output for resistive bridge sensors. Sens. Actuators A **138**(1), 112–119 (2007)

[JUS09]   Julsereewong, A.: Differential voltage-to-frequency converter for telemetry. In: Proceedings of the International MultiConference of Engineers and Computer Scientists (IMECS), vol. 2, pp. 1459–1462. Hong Kong, 18–20 March 2009 (2009)

[McD98]   McDonagh, D., Arshak, K.I.: CMOS bridge to frequency converter with gain and offset control. Microelectronics J. **29**(10), 727–732 (1998)

[PET10]   Petchmaneelumka, W., Julsereewong, A.: Enhanced differential voltage-to-frequency converter for telemetry applications. In: Proceedings of SICE Anual Conference 2010, pp. 3155–3158. Taipei, 18–21 Aug 2010 (2010)

[THE10a]   Thepmanee, T., Julsereewong, A., Sangsuwan, T.: Improved differential voltage-to-frequency converter for telemetry. In: Proceedings of the International Conference on Electrical Engineering/Electronis computer Telecomunications and Information Techonology (ECTI-CON), vol. 1, pp. 216–220. Chiang Mai, 19–21 May 2010 (2010)

[THE10b]   Thepmanee, T., Julsereewong, A.: Simple and low-cost voltage-controlled oscillator. In: Proceedings of the International Conference on Control, Automation and Systems, pp. 1395–1398. Gyeonggi-do, 27–30 Oct 2010 (2010)

[VAL11]   Valero, M.R., Celma, S., Calvo, B., Medrano, N.: CMOS voltage-to-frequency converter with temperature drift compensation, IEEE Trans. Instrum. Meas. **60**(9), 3232–3234 (2011)

[WAN06]   Wang, C.C., Lee, T.J., Li, C.C., Hu, R.: An all-MOS high-linearity voltage-to-frequency converter chip with 520-kHz/V sensitivity. IEEE Trans. Circuits Syst. II **53**(8), 744–747 (2006)

[WAN07]   Wang, C.C., Lee, T.J., Li, C.C., Hu, R.: Voltage to frequency converter with high sensitivity using all-MOS voltage window comparator. Microelectronics J. **38**(2), 197–202 (2007)

Chapter 5
Conclusions

Throughout this book, the most relevant results and main conclusions have been summarized in the concluding discussion of each chapter. In this final chapter, the most significant contributions will be reported to give a general overview of the work.

First, the fulfillment of the main objectives presented in Sect. 1.3 will be verified, leading to the corresponding conclusions. Then, further research directions will be pointed out. Among these are issues not considered in this book, as well as the more in-depth development of some of those already accomplished. These proposed investigations could well be used in future works as an extension to complete the work presented here.

5.1 General Conclusions

The increased emergence of WSN technology has raised the use of the so-called smart sensors to monitor different physical or chemical parameters. These sensors include in the same chip, besides the sensing element, the conditioning circuit that provides a digital signal to be read by a digital port of a microcontroller. Nevertheless, to achieve a low-cost solution, the use of low-cost analog sensors, followed by an only electronic interface integrated in CMOS technology is becoming the preferable choice.

These interfaces have to be low-power and reconfigurable to adapt different types of sensors. They usually consist of a programmable voltage adapter that adjusts the gain and offset of the sensor signals to fit a common output range, and an analog-to-digital converter that digitizes the signal that will be driven to the microcontroller. For digitalization, quasi-digital converters are becoming widely used due to their time-measurement capabilities. Under this premise, this book focused on the design of CMOS voltage-to-frequency converters for this

C.A. Murillo et al., *Voltage-to-Frequency Converters: CMOS Design and Implementation*,   113
Analog Circuits and Signal Processing, DOI 10.1007/978-1-4614-6237-8_5,
© Springer Science+Business Media New York 2013

task because of their overwhelming advantages in terms of simplicity or noise immunity. The essential challenges to cope with are the power consumption, for its use in battery-operated interfaces, and the input range, that must be rail-to-rail for an optimum digitalization.

From an exhaustive review of the most frequently used types of VFCs, multivibrator VFC has been selected for a CMOS low-voltage single-supply implementation attaining low-power operation. This type of VFC usually consists of a voltage-to-current converter, a bidirectional current integrator, and a control circuit.

Several V-I converters have been studied, starting from the classical OTA-NMOS-grounded resistor structure, which achieves high linearity, but exhibits rather limited input range. Though significant swing extension is achieved using an OTA/common source amplifier V-I scheme, this converter still presents a limited input range. Therefore, three compact and simple novel CMOS V-I converters have been presented. These 1.2 V–0.18 μm CMOS converters attain rail-to-rail operation, require low power (below 80 μW), are temperature compensated over a (−40, +120 °C) range and achieve high linearity (maximum transconductance deviation below 30 % over all the temperature range) with an active area below 0.0145 μm^2, showing very competitive performances when compared with the state-of-the-art rail-to-rail V-I converters, being not only suitable for use in VFCs, but also in other applications.

Two proposals for bidirectional current integrators with grounded capacitor compatible with low-voltage low-power design have been presented. The first one is a conventional approach and the second one is less power demanding, allowing a better current copy. The possibility to add programmability and an offset frequency to the VFC by introducing simple modifications in the bidirectional current integrator has been shown. Programmability has been achieved by means of including several digitally controlled current mirrors with different scaling factors, whereas the offset frequency is attained by introducing an offset current at this point, which is added to the scaled current that flows through the current mirrors. With these two features a versatile VFC can be obtained.

The control circuit, responsible for providing the switching signals required by the bidirectional current integrator as well as the VFC output signal, is based on a voltage-window comparator (VWC), consisting of two differential pairs followed by inverters and a $\overline{RS}$ flipflop, instead of a Schmitt trigger comparator, to attain process independent comparison limits and providing a relatively fast response. A power consumption reduction technique is performed, making the comparators in the VWC work alternately. In this block a compromise between power consumption and introduced delay has to be taken into account.

Several proposals to generate the comparison limits of the VWC and the bias current have also been introduced, from the simplest solutions to complete temperature and supply voltage independent solutions. In addition, whenever possible, an enable terminal has been included to set the VFC into a sleep-mode when no conversion is required, thus achieving extreme reduction in the power consumption of the system.

Subsequently, four different VFCs implemented in the 0.18-μm CMOS technology provided by UMC have been developed with the previously presented building blocks.

The first proposal (VFC1) is based on an enhanced V-I converter followed by the conventional bidirectional current integrator, and includes temperature compensation, digitally programmable output frequency range, and a sleep-mode terminal to power down the device when it is inactive. The measured results of this low-voltage ($V_{DD} = 1.8$ V) low-power (0.4 mW in full operation and 35 nW in sleep-mode) VFC confirm an input range of [0.1, 1.6 V] and digitally selectable sensitivities of 312.5, 625 and 937.5 kHz/V with a linearity error of 11 bits over a temperature variation of $(-40, +120\ ^\circ\mathrm{C})$.

The second proposal (VFC2) is based on a rail-to-rail feedback voltage attenuation V-I converter, followed by the low-power bidirectional current integrator, and includes temperature compensation and an offset frequency. The measured results of this low-voltage ($V_{DD} = 1.2$ V) low-power (below 70 μW) VFC confirm a full input range and an output frequency range of [0.1, 1.0 MHz] with a linearity error of 12 bits over a temperature variation of $(-40, +120\ ^\circ\mathrm{C})$.

Finally, two rail-to-rail temperature-compensated differential VFCs (VFC3 and VFC4) based on the rail-to-rail feedback voltage attenuation V-I converter have been proposed. VFC3 is a low-voltage ($V_{DD} = 1.2$ V) low-power (65 μW) dVFC with an output range of [0.0 – 0.9 MHz], a linearity error of 13 bits over temperature variation of $(-40, +120\ ^\circ\mathrm{C})$ and supply voltage variation of ± 30 %. VFC4 is a low-voltage ($V_{DD} = 1.2$ V) low-power (75 μW) dVFC that can operate either with single or differential input signals. Simulation results over a $(-40, +120\ ^\circ\mathrm{C})$ temperature range and supply voltage variations of ± 30 % show an output frequency range of [0.0 – 0.9 MHz] with a linearity error of 13 bits in single mode and a linearity error of 14 bits in differential mode.

The general specifications imposed for the VFCs presented in this book were a maximum voltage supply of $V_{DD} = 1.8$ V, a power consumption below 1 mW, and an output frequency below 2 MHz with temperature and supply voltage independent sensitivity and a maximum linearity error of 0.1 % (10 bits). These requirements have been progressively achieved and, on the whole, the proposed VFCs are more than competitive with those already presented in the literature. They exhibit a larger operating range, coming to obtain rail-to-rail operation, while keeping fairly good performances in terms of sensitivity, linearity and power consumption. Consequently, VFCs are a good alternative for future multisensory read out interfaces to be used in battery-operated systems.

5.2 Further Research Directions

All the proposed VFCs use simple differential pairs followed by inverters as comparators to implement the VWC control circuit, to have a fast response. However, to keep power consumption bounded, the differential pairs biasing current is bounded, causing an increase in the delay introduced to the VFC, thus increasing the relative and sensitivity errors in the system. Therefore, the design of new comparators that have a fast response as well as low-power consumption offers a new research line that could be very advantageous for portable applications.

The β-multiplier bias circuit explained in Sect. 3.4 and the generation circuit of the VWC comparison limits explained in Sect. 3.3, with two resistors driven by the bias current, provide temperature and supply voltage independent bias current and V_H and V_L, respectively. Though these solutions have been validated through the operation of VFC3, and VFC4, the research in the design of bias circuits and voltage references introduces a new field of study that should be developed further to obtain optimum LV-LP solutions, and has been set aside for future investigation lines. Another open issue in the field of biasing is the implementation of an on-chip voltage regulator that feeds the complete system, thus having full V_{DD} insensitivity. This voltage regulator should include an easy enable that drives the system into a sleep-mode, to achieve a complete on-chip solution that includes power management.

Finally, integration of the programmable voltage adapter in the same chip of the VFC is desirable to have a complete electronic multisensor interface.

Thus, the future work clearly outlined will aid in the completion of the study carried out in this book as well as help to explore new research direction.

Appendix A: UMC 0.18-µm Mixed-Mode/RF CMOS Process

UMC L180 MM/RF 1.8 V/3.3 V 1P6M technology is the CMOS process based on general P-Sub structure with one layer of poly, six layers of aluminum metal and FSG dielectrics. Moreover, the MM/RF process also includes several optional layers which defined and decided by customer's applications design (Tables A.1 and A.2).

Key process features

- P-Sub CMOS process with optional Deep N-well.
- Dual gate oxide thickness (1.8 V/3.3 V).
- Mixed optional device application is available.
- Three M6 aluminum thickness types depend on customer's design application.
- FSG dielectrics.

C.A. Murillo et al., *Voltage-to-Frequency Converters: CMOS Design and Implementation*,   117
Analog Circuits and Signal Processing, DOI 10.1007/978-1-4614-6237-8,
© Springer Science+Business Media New York 2013

Table A.1 Key device parameters

Device type	Core		I/O	
Parameter	RVT	LVT	RVT	LVT
V_{CC} (V)	1.8		3.3	
t_{OX} (Å)	33		65	
L_{min_draw} (μm)	0.18		0.34	
V_{TH} N/P (V)	0.51/−0.5	0.22/−0.22	0.65/−0.7	0.31/−0.42
I_{DS} N/P (μA/μm)	625/−244.2	720/−270	590/−260	640/−250
I_{OFF} N/P (A/μm)	7.6p/−8.1p	29.4n/−12.4n	1p/−0.5p	0.92n/−2p
Gate delay (ps/stage)	28.5	−	55	−

Table A.2 Key design rules

Layers	Min. width (μm)	Min. spacing (μm)	Pitch (μm)
Diffusion	0.24	0.28	0.52
Inter-well	0.9	0.9	1.8
Drawn poly	0.18	0.28	0.46
Contact	0.24	0.26	0.5
Metal 1	0.24	0.24	0.48
MVia1—MVia5	0.28	0.28	0.56
Inter metal	0.28	0.28	0.56
Metal_Cap	0.6	0.55	1.15
Metal 6 8 K	0.44	0.44	0.88
Metal 6 12 K	0.8	0.8	1.6
Metal 6 20 K	1.2	1.0	2.2

Appendix B: V-I Converters: Small Signal Analysis

In this appendix, the detailed frequency response analysis of the voltage-to-current converters presented in Sect. 3.1 is carried out.

These types of feedback systems can be described by different parameters, as shown in Fig. B.1, where x_i is the input signal, x_a is an internal signal, x_0 is the output signal, $C = x_a/x_i$ is the input gain, $K = x_0/x_a$ is the amplifier gain and $D = (x_a/x_0)|_{x_i = 0}$ is the return ratio. The output can be obtained as

$$x_0 = \frac{KC}{1 - KD}x_i \qquad (B.1)$$

Therefore, the poles of the feedback system can be obtained from the characteristic equation $1 - K(s)D(s) = 0$ [ROS74].

B.1 Conventional V-I Converter

The conventional V-I converter, shown in Fig. B.2, consists of an OTA driving a MOS (T_0) and a grounded linear resistor (R_S) in a negative feedback loop. For its analysis, the OTA is going to be described with a transconductance G and a dominant pole form by the output capacitance C and resistance R, following the scheme shown in Fig. B.3. The source follower fits the scheme shown in Fig. B.4, where a simplified small-signal transistor model has been used and parameters denote their usual meaning, using the subscript $_0$ for transistor T_0.

In the mid frequency range, the source follower gain is approximately given by (B.2), where R_L is the load resistor which is the parallel connection of R_S and $1/g_{mb0}$.

$$\left.\frac{V_A}{V_g}\right|_{\omega \to 0} = \frac{g_{m0}R_L}{1 + g_{m0}R_L} \qquad (B.2)$$

C.A. Murillo et al., *Voltage-to-Frequency Converters: CMOS Design and Implementation*, 119
Analog Circuits and Signal Processing, DOI 10.1007/978-1-4614-6237-8,
© Springer Science+Business Media New York 2013

Fig. B.1 Block diagram
of the feedback system

Fig. B.2 Schematics of the
conventional V-I converter

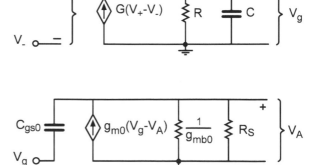

Fig. B.3 Equivalent linear
model of the OTA

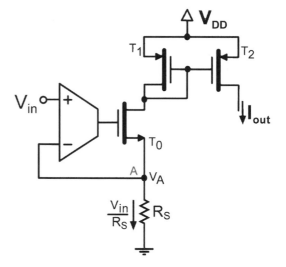

Fig. B.4 Equivalent linear
model of the source follower

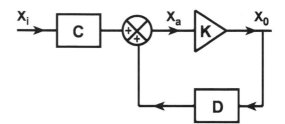

Fig. B.5 Linear model of the system to obtain $D(s)$

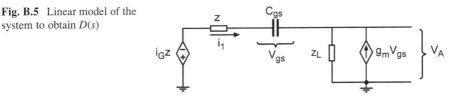

Analyzing the OTA, where $V_+ = V_{in}$ and $V_- = V_A$, the gate voltage V_g is given by

$$V_g|_{\omega \to 0} = G(V_{in} - V_A)R \tag{B.3}$$

Thus, the voltage V_A can be expressed as

$$\left.\frac{V_A}{V_{in}}\right|_{\omega \to 0} = \frac{GRg_{m0}R_L}{1 + g_{m0}R_L + g_{m0}R_LGR} \approx \frac{GR}{1 + GR} \approx 1 \tag{B.4}$$

Now, for the frequency response, let us assume that $x_i = V_{in}$ and $x_0 = V_A$. Choosing $K(s) = i_G/V_A = G$ and $D(s) = V_A/i_G$, as shown in Fig. B.5, from the analysis of the system, three equations can be obtained, considering z the parallel connection of C and R.

$$i_G z + i_1 z + V_{gs0} + V_A = 0 \tag{B.5}$$

$$i_1 = sC_{gs0}V_{gs0} \tag{B.6}$$

$$i_1 + g_{m0}V_{gs0} = \frac{V_A}{R_L} \tag{B.7}$$

From (B.5)–(B.7), $D(s)$ is given by

$$D(s) = \frac{-zR_L(g_{m0} + sC_{gs0})}{(g_{m0} + sC_{gs0})R_L + (1 + szC_{gs0})} \tag{B.8}$$

Now, substituting z by $R/(1 + sRC)$, the characteristic equation, $1 - K(s)D(s) = 0$, is approximated by

$$s^2CC_{gs0} + s(Cg_{m0} + GC_{gs0}) + g_{m0}G = 0 \tag{B.9}$$

Therefore, the transfer function is given by

$$\frac{V_A}{V_{in}}(s) = \frac{GR}{1 + GR} \frac{g_{m0}G}{s^2CC_{gs0} + s(Cg_{m0} + GC_{gs0}) + g_{m0}G} \tag{B.10}$$

And the poles of the system can be extracted from (B.11).

$$s^2 + s\left(\frac{g_{mo}}{C_{gs0}} + \frac{G}{C}\right) + \frac{g_{m0}G}{CC_{gs0}} = s^2 - (p_1 + p_2)s + p_1 p_2 = 0 \qquad (B.11)$$

If a dominant pole exists, $|p_2| \gg |p_1|$, the following approximation can be done

$$p_1 + p_2 \approx p_2 \qquad (B.12)$$

Obtaining that the poles are

$$p_2 = -\frac{g_{m0}}{C_{gs0}} - \frac{G}{C} = -\frac{Cg_{m0} + GC_{gs0}}{C_{gs0}C} \qquad (B.13)$$

$$p_1 = -\frac{C_{gs0}C}{Cg_{m0} + GC_{gs0}}\frac{g_{m0}G}{CC_{gs0}} = -\frac{g_{m0}G}{Cg_{m0} + GC_{gs0}} \qquad (B.14)$$

where it is easy to see that $|p_2| \gg |p_1|$. These two poles fits the intuitive idea that $p_1 = -g_{m0}/C_{gs0}$ and $p_2 = -G/C$. So, the voltage V_{in} is buffered to V_A according to the expression:

$$\frac{V_A}{V_{in}}(s) = \frac{GR}{1 + GR}\frac{g_{m0}G}{s(Cg_{m0} + GC_{gs0}) + g_{m0}G}\frac{Cg_{m0} + GC_{gs0}}{C_{gs0}Cs + (Cg_{m0} + GC_{gs0})} \qquad (B.15)$$

Making the approximation of dominant pole, and taking into account that $GR/(1 + GR) \approx 1$, the expression of the voltage V_{in} is given by

$$\frac{V_A}{V_{in}}(s) \approx \frac{g_{m0}G}{s(Cg_{m0} + GC_{gs0}) + g_{m0}G} \qquad (B.16)$$

To verify that the obtained response is valid, an electric simulation has been carried out, as shown in Fig. B.6, with the OTA of two complementary pairs used in rail-to-rail V-I converters in Chap. 3, a resistor $R_S = 40$ kΩ, a NMOS transistor $T_0 = (200$ μm/0.5 μm$)$, and PMOS transistors in the mirror with sizes $T_1 = T_2 = (300$ μm/1.2 μm$)$.

If a simple mirror is used, the transconductance of the conventional converter is

$$G_M(s) = \frac{V_A(s)}{R_S V_{in}}\frac{(W/L)_2}{(W/L)_1}\frac{g_{m1}}{g_{m1} + s(C_{gs1} + C_{gs2})}(1 + \varepsilon) \qquad (B.17)$$

where ε is the deviation of the current copy from its ideal value, due to unequal T_1 and T_2 drain-source voltages and channel length modulation. If a high-swing cascode current mirror is used to reduced the deviation, the transconductance is given by

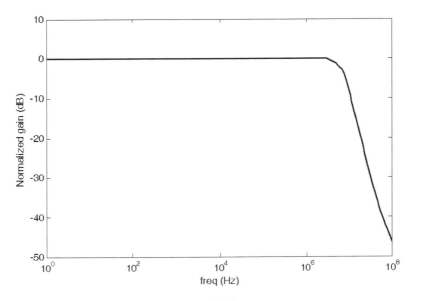

Fig. B.6 Frequency response of the conventional V-I converter

$$G_M(s) = \frac{V_A(s)}{R_S V_{in}} \frac{(W/L)_2}{(W/L)_1} \frac{1}{\dfrac{C_{gs1} + C_{gs2}}{g_{m1}g_{m2}} C_{gs2}s^2 + \dfrac{C_{gs1} + C_{gs2}}{g_{m1}} g_{m2}s + 1} \qquad (B.18)$$

B.2   Enhanced V-I Converter

The enhanced V-I converter is based on an OTA/common-source amplifier config-uration, as shown in Fig. B.7. For its analysis, the OTA is modeled by a transcon-ductance G, and a dominant pole formed by an output capacitance C and resistor R, following the scheme shown in Fig. B.3. The cascode common-source equivalent linear scheme is shown in Fig. B.8, where a simplified scheme has been used to modeling the transistors, and only capacitances that contributes with a significant time constant have been considered. Subscripts $_1$ and $_{1C}$ denote transistor T_1 and T_{1C} respectively.

For this analysis, as the voltage buffering and the current copy are coupled, both the buffered voltage V_A and the output current I_{out} are going to be obtained at the same time. Cascode transistor are chosen to be $T_{1C} = T_1$ and $T_{2C} = T_2$, therefore, $r_{01} = r_{01C}$ and $r_{02} = r_{02C}$.

First, an analysis in the mid frequency range is carried out. Thus, V_C/V_g can obtained, taking into account that T_{1C} contributes like two parallel resistors of values $1/g_{m1C}$ and $1/g_{mb1}$.

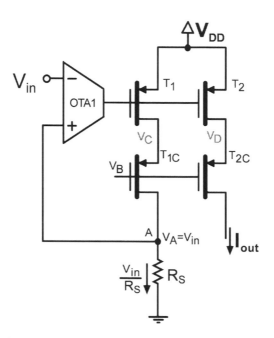

Fig. B.7 Enhanced V-I converter

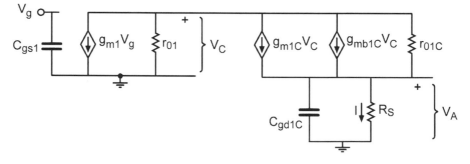

Fig. B.8 Equivalent linear model of the cascode common-source

$$\frac{V_C}{V_g}\bigg|_{\omega \to 0} = -g_{m1}\frac{r_{01}}{1 + r_{01}(g_{m1C} + g_{mb1C})} \tag{B.19}$$

where parameters have their usual meaning.

Then V_A/V_C is calculated.

$$\frac{V_A}{V_C}\bigg|_{\omega \to 0} = \frac{R_S}{R_S + r_{01}}[1 + (g_{m1C} + g_{mb1C})r_{01}] \tag{B.20}$$

Voltage V_A across resistor R_S can also be given as a function of V_g, obtained from (B.19) and (B.20).

$$\left.\frac{V_A}{V_g}\right|_{\omega\to 0} = -g_{m1}(r_{01}\|R_S) \tag{B.21}$$

Next, V_g/V_{in} is calculated as $V_g = GR(V_A - V_{in})$, where V_A is the one given in (B.21).

$$\left.\frac{V_g}{V_{in}}\right|_{\omega\to 0} = -\frac{GR}{1 + GR(r_{01}\|R_S)g_{m1}} \tag{B.22}$$

The voltage V_A is finally obtained, and it is given by

$$\left.\frac{V_A}{V_{in}}\right|_{\omega\to 0} = \frac{g_{m1}(r_{01}\|R_S)GR}{1 + GR(r_{01}\|R_S)g_{m1}} \approx 1 \tag{B.23}$$

And the transconductance G_M is given by

$$\begin{aligned}
G_M|_{\omega\to 0} &= -g_{m2}\frac{V_g}{V_{in}} = g_{m2}\frac{GR}{1 + GR(r_{01}\|R_S)g_{m1}} \\
&\approx \frac{g_{m2}}{g_{m1}}\frac{1}{(r_{01}\|R_S)} \approx \frac{(W/L)_2}{(W/L)_1}\frac{1}{R_S}
\end{aligned} \tag{B.24}$$

For the study of the frequency response, K and D are chosen $K(s) = i_G/V_A = G$ and $D(s) = V_A/i_G$ as shown in Fig. B.9, where R_L is the load resistor, and the cascode common-source T_2–T_{2C} follows the scheme shown in Fig. B.8.

Analyzing the system, the following three equations can be obtained

$$V_g(s) = i_G\frac{R}{1 + sRC_T} \tag{B.25}$$

$$\frac{V_C}{V_g}(s) = -\frac{g_{m1}r_{01}}{1 + r_0(g_{m1C} + g_{mb1C})} \tag{B.26}$$

$$\frac{V_A}{V_C}(s) = \frac{R_S}{R_S + r_{01}(1 + sR_SC_{gd1C})}[1 + (g_{m1C} + g_{mb1C})r_{01}] \tag{B.27}$$

where $C_T = C + C_{gs1} + C_{gs2}$. From (B.25)–(B.27), the expression for $D(s)$ is given by

$$D(s) = -g_{m1}r_{01}\frac{R}{1 + sRC_T}\frac{R_S}{R_S + r_{01}(1 + sR_SC_{gd1C})} \tag{B.28}$$

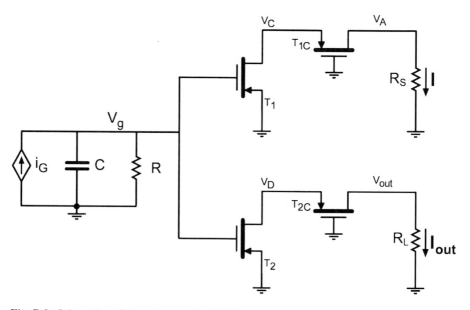

Fig. B.9 Schematics of the system to obtain $D(s)$

It can be approximated, assuming that $R_S \ll r_{01}$, by

$$D(s) \approx -\frac{R}{1 + sRC_T}\frac{g_{m1}R_s}{1 + sR_SC_{gd1C}} \tag{B.29}$$

Making use of (B.22), the gate voltage can be expressed as

$$\frac{V_g}{V_{in}}(s) = \frac{-GR}{1 + GRR_Sg_{m1}} \cdot$$

$$\frac{1 + g_{m1}R_SGR}{s^2(RR_SC_TC_{gd1C}) + s(R_SC_{gd1C} + RC_T) + (1 + g_{m1}R_SGR)} \tag{B.30}$$

And the voltage V_A across resistor R_S, obtained from (B.30) using (B.26) and (B.27), can be expressed as

$$\frac{V_A}{V_{in}}(s) = \frac{-R_Sg_{m1}r_{01}}{R_S + r_{01}(1 + sR_SC_{gd1C})}\frac{V_g}{V_{in}} \tag{B.31}$$

With the approximations of $R_S \ll r_0$ and $1 \ll GRR_Sg_{m1}$, and substituting V_g/V_{in}, (B.30) remains as follows

$$\frac{V_A}{V_{in}}(s) = \frac{1}{(1 + sR_S C_{gd1C})} \cdot$$
$$\frac{1 + g_{m1}R_S GR}{s^2(RR_S C_T C_{gd1C}) + s(R_S C_{gd1C} + RC_T) + (1 + g_{m1}R_S GR)} \quad \text{(B.32)}$$

Now, the output current transconductance G_M is determined. First, V_D/V_g is obtained

$$\frac{V_D}{V_g}(s) = -\frac{g_{m2}r_{01}}{1 + r_{01}(g_{m2C} + g_{mb2C})} \quad \text{(B.33)}$$

Then, the output voltage V_{out}

$$\frac{V_{out}}{V_D}(s) = \frac{R_L}{R_L + r_0(1 + sR_L C_{gd2C})}[1 + (g_{m2C} + g_{mb2C})r_{01}] \quad \text{(B.34)}$$

And therefore transconductance is given by

$$G_M(s) = \frac{-g_{m2}r_{01}}{R_L + r_{01}(1 + sR_L C_{gd2})} \frac{V_g}{V_{in}} \approx \frac{-g_{m2}}{(1 + sR_L C_{gd2C})} \frac{V_g}{V_{in}} \quad \text{(B.35)}$$

Substituting V_g/V_{in} and doing the approximations of $R_S \ll r_{01}$ and $1 \ll GRR_S g_{m1}$, the expression of G_M is given by

$$G_M(s) = \frac{g_{m2}}{R_S g_{m1}} \frac{1}{(1 + sR_L C_{gd2C})} \cdot$$
$$\frac{1 + g_{m1}R_S GR}{s^2(RR_S C_T C_{gd1C}C) + s(R_S C_{gd1C} + RC_T) + (1 + g_{m1}R_S GR)} \quad \text{(B.36)}$$

The expression (B.36) presents three poles, p_1, p_2 and p_3. One of them is $p_1 = -1/R_L C_{gd2C}$ but the other two, which also appear in the V_A expression, are coupled. Therefore, they are going to be extracted from (B.37).

$$s^2 + s\left(\frac{1}{R_0 C_T} + \frac{1}{R_S C_{gd1C}}\right) + \frac{g_{m1}G}{C_T C_{gd1C}} = s^2 - (p_2 + p_3)s + p_2 p_3 = 0 \quad \text{(B.37)}$$

Assuming a dominant pole behavior, $|p_3| \gg |p_2|$, the following approximation can be done.

$$p_2 + p_3 \approx p_3 \quad \text{(B.38)}$$

Obtaining that the poles are given by (B.39) and (B.40).

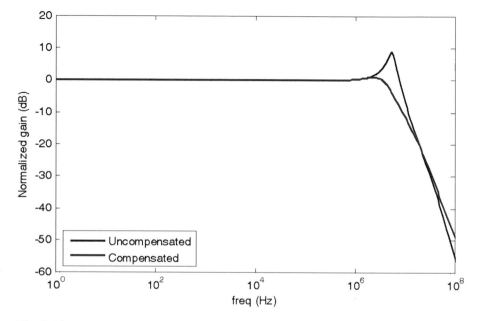

Fig. B.10 Frequency response of the enhanced V-I converter with and without compensation network

$$p_3 = -\frac{1}{RC_T} - \frac{1}{R_S C_{gd1C}} = -\frac{R_S C_{gd1} + RC_T}{RC_T R_S C_{gd1C}} \tag{B.39}$$

$$p_2 = -\frac{g_{m1}G}{C_T C_{gd1C}} \frac{RC_T R_S C_{gd1C}}{R_S C_{gd1C} + RC_T} = -\frac{g_{m1}GRR_S}{R_S C_{gd1C} + RC_T} \tag{B.40}$$

Looking to the obtained poles, we cannot assure that the assumption of dominant pole was correct. This fits with the frequency response shown in Fig. B.10 in black, where the V-I converter has been simulated with the OTA of two complementary pairs used in rail-to-rail V-I converters in Chap. 3, a resistor $R_S = 40$ kΩ and $T_1 = T_2 = (300\ \mu m/1.2\ \mu m)$. Therefore, in order to avoid undesirable peaks in the closed loop frequency response and underdamped oscillations, compensation is included. As it is known from the theory, when adding a compensation capacitor C_C, the poles of the system are split (and remain in the left semi-plane). However, it also introduces a zero in the frequency response, which is in the right semi-plane, worsening the phase margin. To avoid this zero being near the unity-gain frequency, a compensation resistor R_C is introduced. This resistor hardly affects the system poles, and it is not going to be considered in the analysis. Thus, a $R_C C_C$ compensation network is introduced in the OTA/common-source amplifier, which can be seen as a conventional Miller compensation in a two stage operational amplifier [BAK98].

The effect of this compensation is now studied, included in C'_T, where $C'_T = C_T + C_C$, without affect the first pole p_1. Thus, the three poles are given by (B.41)–(B.43), where the assumption of $RC'_T \gg R_S C_{gd1C}$ has been done.

$$p_3 = -\frac{1}{RC'_T} - \frac{1}{R_S C_{gd1C}} = -\frac{R_S C_{gd1} + RC'_T}{RC'_T R_S C_{gd1C}} \approx -\frac{1}{R_S C_{gd1C}} \tag{B.41}$$

$$p_2 = -\frac{g_{m1}G}{C'_T C_{gd1C}} \frac{RC'_T R_S C_{gd1C}}{R_S C_{gd1C} + RC'_T} = -\frac{g_{m1}GRR_S}{R_S C_{gd1} + RC'_T} \approx -\frac{g_{m1}GR_S}{C'_T} \tag{B.42}$$

$$p_1 = -\frac{1}{R_L C_{gd2C}} \tag{B.43}$$

From the above equations, it is shown that $|p_3| \gg |p_2|$ and $|p_1| \gg |p_2|$. Thus, making the approximation of a dominant pole behavior, the voltage V_A and the transconductance G_M are given by the expressions

$$\frac{V_A}{V_{in}}(s) \approx \frac{g_{m1}GR_S}{(g_{m1}GR_S + sC'_T)} \tag{B.44}$$

$$G_M(s) \approx \frac{1}{R_S} \frac{g_{m2}}{g_{m1}} \frac{g_{m1}GR_S}{(g_{m1}GR_S + sC'_T)} \tag{B.45}$$

A new simulation has been carried out, adding a compensation network ($R_C = 1$ kΩ, $C_C = 0.4$ pF) to the converter, shown in Fig. B.10 in red. Thanks to the compensation network no peaks appear in the system response.

B.3 FFVA V-I Converter

The feedforward voltage attenuation (FFVA) V-I converter is shown in Fig. B.11. This converter basically consists of two cascaded enhanced V-I converters, where the input of the second stage has been attenuated by means of a resistive voltage divider formed by linear resistors R_1 and R_2. So, the attenuator is made by an auxiliary OTA/common-source formed by OTA_{aux}–T_{A1}–T_{A1C}–R_{CA}–C_{CA} and R_1 and R_2.

The equations obtained for the enhanced V-I converter can be directly applied to the system. Thus, the input voltage V_{in} is buffered to V_{out1} by the OTA_{aux}–T_{A1}–T_{A1C} voltage follower with rail-to-rail input–output operation. According to (B.23), the buffered voltage is given by

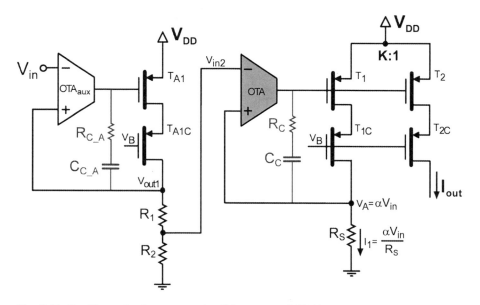

Fig. B.11 Feedforward voltage attenuation V-I converter (FFVA)

$$\left.\frac{V_{out1}}{V_{in}}\right|_{\omega \to 0} = \frac{g_{m1\_A}\left(r_{01\_A}||(R_1 + R_2)\right)G_{\_A}R_{\_A}}{1 + G_{\_A}R_{\_A}g_{m1\_A}\left(r_{01\_A}||(R_1 + R_2)\right)} \approx 1 \qquad (B.46)$$

where the subscript $_{\_A}$ denotes the parameters of the auxiliary elements OTA_{aux}–T_{A1}–T_{AIC}. Then, the buffered voltage V_{out1} is attenuated to $V_{in2} = \alpha V_{out1}$ through the voltage divider formed by resistors R_1 and R_2, with α given by

$$\alpha = \frac{R_2}{R_1 + R_2} \qquad (B.47)$$

This attenuated voltage αV_{in} is the input of the main V-I converter formed by OTA–T_1–T_{1C} and resistor R_S. Thus, αV_{in} is followed to V_A, according to

$$\left.\frac{V_{out1}}{V_{in2}}\right|_{\omega \to 0} = \frac{g_m(r_{01}||R_S)GR}{1 + GRg_m(r_{01}||R_S)} \approx 1 \qquad (B.48)$$

From the output current expression obtained in (B.24), the transconductance of the whole FFVA V-I converter is given by

$$\left.G_M\right|_{\omega \to 0} \approx \alpha \frac{(W/L)_2}{(W/L)_1}\frac{1}{R_S} \qquad (B.49)$$

Following the same analysis, the frequency response can be obtained from (B.44) and (B.45). Thus, the voltage V_A across resistor R_S is

$$\frac{V_A}{V_{in}}(s) \approx \alpha \frac{GBW}{(GBW + s)} \frac{GBW_A}{(GBW_A + s)} \tag{B.50}$$

And the transconductance G_M is therefore given by

$$G_M(s) = \frac{\alpha}{R_S} \frac{(W/L)_2}{(W/L)_1} \frac{GBW}{(GBW + s)} \frac{GBW_A}{(GBW_A + s)} \tag{B.51}$$

being $GBW = g_{m1}GR_S/C'_T$ and $GBW_A = g_{m1\_A}G_A(R_1 + R_2)/C'_{T\_A}$, and the capacitances $C'_T = C + C_{gs1} + C_{gs2} + C_C$ and $C'_{T\_A} = C_A + C_{gs1\_A} + C_{C\_A}$.

B.4   FBVA V-I Converter

The feedback voltage attenuation (FBVA) V-I converter, shown in Fig. B.12, is based on the enhanced V-I converter, but it includes a floating dynamic battery introduced between the OTA non-inverting input and node A. It is implemented by using a non-inverting amplifier stage formed by OTA$_{aux}$–T$_{A1}$–T$_{A1C}$ and feedback resistors R_1 and R_2. For its analysis, the OTAs are described by a transconductance G, and a dominant pole formed by the output capacitance C and resistance R, in the case of the main OTA and G_A, C_A and R_A in the auxiliary one, following the scheme shown in Fig. B.7. In the following, subscript $_A$ denotes that the parameter belongs to

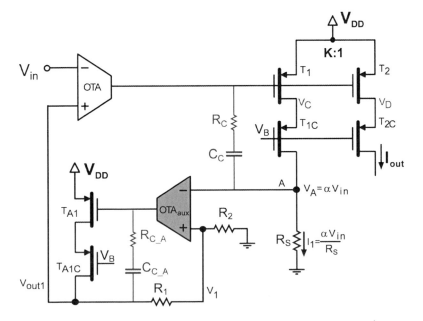

Fig. B.12 Feedback voltage attenuation V-I converter (FBVA)

an auxiliary element. Transistors of the common-source follow the scheme shown in Fig. B.8.

Analyzing the system, in the mid-frequency range, it can be described by the following expressions

$$V_g\big|_{\omega\to0} = G(V_{out1} - V_{in})R \tag{B.52}$$

$$\frac{V_C}{V_g} = -\frac{g_{m1}r_{01}}{1 + (g_{m1C} + g_{mb1C})r_{01}} \tag{B.53}$$

$$\frac{V_A}{V_C} = \frac{R_S[1 + (g_{m1C} + g_{mb1C})r_{01}]}{R_S + r_{01}} \tag{B.54}$$

$$V_{g\_A}\big|_{\omega\to0} = G_{\_A}(V_1 - V_A)R_{\_A} \tag{B.55}$$

$$\frac{V_1}{V_{out1}} = \alpha = \frac{R_2}{R_1 + R_2} \tag{B.56}$$

$$\frac{V_D}{V_{g\_A}} = -\frac{g_{m1\_A}r_{01\_A}}{1 + (g_{m1C\_A} + g_{mb1C\_A})r_{01\_A}} \tag{B.57}$$

$$\frac{V_{out1}}{V_D} = \frac{(R_1 + R_2)[1 + (g_{m1C\_A} + g_{mb1C\_A})r_{01\_A}]}{R_1 + R_2 + r_{01\_A}} \tag{B.58}$$

Next, V_{out}/V_A is calculated from the above (B.55)–(B.58).

$$\frac{V_{out1}}{V_A}\bigg|_{\omega\to0} = \frac{((R_1 + R_2)\|r_{01\_A})g_{m1\_A}G_{\_A}R_{\_A}}{1 + ((R_1 + R_2)\|r_{01\_A})g_{m1\_A}G_{\_A}R_{\_A}\alpha} \approx \frac{1}{\alpha} \tag{B.59}$$

And then V_A/V_{in} is obtained from (B.52)–(B.54), and (B.59), and it is given by

$$\frac{V_A}{V_{in}}\bigg|_{\omega\to0} = \frac{(r_{01}\|R_S)g_{m1}GR}{1 + (r_{01}\|R_S)g_{m1}GR\frac{1}{\alpha}} \approx \alpha \tag{B.60}$$

And transconductance G_M is expressed as (B.61), where V_g/V_{in} has been obtained from (B.53), (B.54), and (B.60).

$$G_M\big|_{\omega\to0} = -g_{m2}\frac{V_g}{V_{in}} = \frac{g_{m2}GR}{1 + (r_{01}\|R_S)g_{m1}GR\frac{1}{\alpha}} \approx \frac{\alpha}{R_S}\frac{g_{m2}}{g_{m1}} \tag{B.61}$$

As demonstrated for in the enhanced V-I converter, the OTA-common source can be approximated by a dominant pole if a compensation network is included. Thus, to make easier the analysis, the rest of the poles are going to be neglected because a compensation network has been used in both, the auxiliary and the main OTA/common-source amplifiers. For the frequency response, the above equations are valid, but instead of (B.52) and (B.55), two new equations are introduced, where $C'_T = C + C_C + C_{gs1} + C_{gs2}$, and $C'_{T\_A} = C_A + C_{C\_A} + C_{gs1\_A} + C_{gs2\_A}$.

$$V_g(s) = G(V_{out1} - V_{in}) \frac{R}{1 + sRC'_T} \tag{B.62}$$

$$V_g(s) = G(V_{out1} - V_{in}) \frac{R}{1 + sRC'_T} \tag{B.63}$$

Following the same analysis made for the mid-frequency range, the attenuated voltage V_A can be approximated by

$$\frac{V_A}{V_{in}}(s) \approx \frac{GBW(\alpha GBW_A + s)}{s^2 + \alpha GBW_A s + GBW_A GBW} \tag{B.64}$$

Where $GBW = G g_{m1}(r_{01}\|R_S)/C'_T$ and $GBW_A = G_A g_{m1\_A}(r_{01\_A}\|(R_1 + R_2))/C'_{T\_A}$ and α is the one introduced in (B.56).
And the transconductance of the FBVA is given by

$$G_M(s) = \frac{I_{out}}{V_{in}} = \frac{V_A}{V_{in}} \frac{g_{m2}}{1 + sR_L C_{gd2C}} \frac{1}{g_{m1}R_S} \tag{B.65}$$

$$G_M(s) \approx \frac{(W/L)_2}{(W/L)_1} \frac{1}{R_S} \frac{GBW(\alpha GBW_A + s)}{s^2 + \alpha GBW_A s + GBW_A GBW} \tag{B.66}$$

B.5   FFCA V-I Converter

The feedforward current attenuation (FFCA) V-I converter is shown in Fig. B.13. It is based on the enhanced V-I converter including a floating dynamic battery between the OTA non-inverting input and node A. The battery is made by using a linear resistor R_2 driven by a current source proportional to V_{in}. To generate the required current, an auxiliary V-I converter formed by OTA$_{aux}$, T$_{A1}$–T$_{A1C}$ and R_1 is used. For its analysis, the OTAs are described by a transconductance G, and a dominant pole formed by the output capacitance C and resistor R, in the case of the main OTA and G_A, C_A and R_A in the auxiliary one, following the scheme shown in Fig. B.7. In the following, subscript $_A$ denotes that the parameter belongs to an auxiliary element. Transistors of the common-source follow the scheme shown in Fig. B.8.

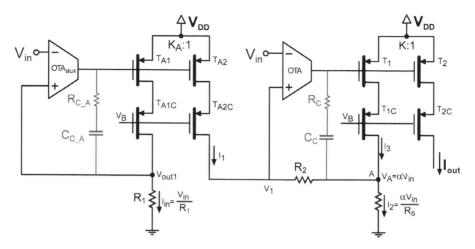

Fig. B.13 Feedforward current attenuation V-I converter (FFCA)

First, the input voltage V_{in} is buffered to V_{out1} by the OTA_{aux}–T_{A1}–T_{A1C} voltage follower with rail-to-rail input–output operation. According to (B.23) the buffered voltage V_{out1} is given by (B.67). The generated current I_{in} across resistor R_1 is replicated through transistors T_{A1}–T_{A1C}–T_{A2}–T_{A2C} with a scaling factor $K_A = (W/L)_{1\_A}/(W/L)_{2\_A}$, and thus, and according to (B.24), the transconductance for the current I_1 is given by (B.68).

$$\left.\frac{V_{out1}}{V_{in}}\right|_{\omega\to0} \approx \frac{g_{m1\_A}G_A(r_{01\_A}\|R_1)R_A}{1 + g_{m1\_A}G_A(r_{01\_A}\|R_1)R_A} \approx 1 \tag{B.67}$$

$$\left.\frac{I_1}{V_{in}}\right|_{\omega\to0} \approx g_{m2\_A}\frac{G_A R_A}{1 + G_A R_A(r_{01\_A}\|R_1)g_{m1\_A}} \approx \frac{(W/L)_{2\_A}}{(W/L)_{1\_A}}\frac{1}{R_1} = \frac{1}{K_A R_1} \tag{B.68}$$

The current I_1 drives a resistor R_2, so that the voltage across R_2 is $V_{R2} = I_1 R_2$, being the positive input of the main OTA $v_+ = V_1 = V_A + I_1 R_2$. To obtain the voltage V_A three equations come up.

$$\frac{V_C}{V_g} = -\frac{g_{m1}r_{01}}{1 + (g_{m1C} + g_{mb1C})r_{01}} \tag{B.69}$$

$$V_g|_{\omega\to0} = G(V_1 - V_{in})R \tag{B.70}$$

$$\frac{V_A}{V_C} = \frac{R_S[1 + (g_{m1C} + g_{mb1C})r_{01}]}{R_S + r_{01}} \tag{B.71}$$

So making use of the above equations and of the expression of I_1 given in (B.68), the ratio V_g/V_{in} is obtained.

$$\left.\frac{V_g}{V_{in}}\right|_{\omega \to 0} = -\frac{g_{m1}(r_{01}\|R_S)RG\alpha}{1 + g_{m1}(r_{01}\|R_S)RG} \tag{B.72}$$

where α is the attenuation factor, which fulfills the following expression

$$\alpha = \frac{K_A R_1 - R_2}{K_A R_1} \tag{B.73}$$

Therefore, the attenuated voltage V_A is given by

$$\left.\frac{V_A}{V_{in}}\right|_{\omega \to 0} = -g_{m1}(r_{01}\|R_S)\frac{V_g}{V_{in}} = \frac{g_{m1}(r_{01}\|R_S)RG\alpha}{1 + g_{m1}(r_{01}\|R_S)RG} \approx \alpha \tag{B.74}$$

For the output current, it has to be taken into account that I_{out} is not the replica of I_2 but of I_3, being $I_3 = I_1 - I_2$, with I_1 given in (B.68) and $I_2 = V_A/R_S$. The replica of I_3, can be obtained from V_{out}/V_g taking into account the load resistor R_L, or can be directly approximated as $I_{out} = I_3 g_{m2}/g_{m1} = I_3(W/L)_2/(W/L)_1$. Thus, with the approximations made above, the transconductance G_M is

$$\left.G_M\right|_{\omega \to 0} = \frac{(W/L)_2}{(W/L)_1}\frac{K_A R_1 - R_2 - R_S}{K_A R_1 R_S} \tag{B.75}$$

To obtain the frequency response of the converter, in the auxiliary part of the converter (B.44) and (B.45) can be applied directly, whereas for the analysis of the second part, the approximation of dominant pole by adding a compensation capacitor C_C is going to be done. In the analysis $C'_T = C + C_C + C_{gs1} + C_{gs2}$. Following the analysis made for mid-frequency range, and using instead of (B.70) the following equation,

$$V_g(s) = G(V_1 - V_{in})\frac{R}{1 + sRC'_T} \tag{B.76}$$

the voltage V_A is given by

$$\frac{V_A}{V_{in}}(s) = \frac{GBW(\alpha GBW_A + s)}{(GBW_A + s)(GBW + s)} \tag{B.77}$$

where $GBW = g_{m1}GR_S/C'_T$, and $GBW_A = g_{m1\_A}G_A R_1/C'_{T\_A}$. And the transconductance of the FFCA is given by

$$G_M(s) = \frac{I_{out}}{V_{in}} = \frac{(W/L)_2}{(W/L)_1} \cdot$$
$$\frac{GBW\,GBW_A(K_A R_1\alpha - R_S) + (K_A R_1 GBW - R_S GBW_A)s}{R_S K_A R_1(GBW_A + s)(GBW + s)} \tag{B.78}$$

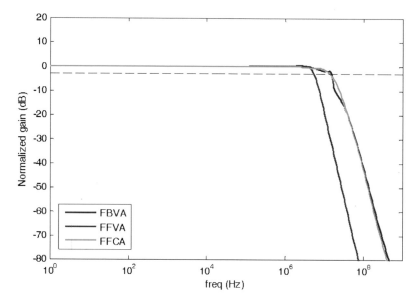

Fig. B.14 Simulated AC response of the FFVA, FBVA and FFCA V-I converters presented in Chap. 3

B.6   Comparison

This brief section makes a comparison between the frequency response of the presented rail-to-rail V-I converters: FFVA, FBVA, and FFCA. Looking to the obtained frequency responses, given by (B.51), (B.66), and (B.78) respectively, it can be seen that FFVA will exhibit a limited bandwidth while a larger bandwidth can be achieved for the FBVA and FFCA because their transfer functions present one zero and two poles, being possible a pole-zero cancelation.

These rail-to-rail converters presented in Chap. 3, have been simulated with a load resistor of $R_L = 1$ kΩ, as shown in Fig. B.14, obtaining bandwidths of $BW_{FFVA} = 5.2$ MHz, $BW_{FBVA} = 14.5$ MHz and $BW_{FFCA} = 14.1$ MHz, validating the above mentioned.

References

[BAK98] Baker, R.J., Li, H.W., Boyee, D.E.: CMOS circuit design, layout & simulation. IEEE press, New York (1997)

[ROS74] Rosenstark, S.: A simplified method of feedback amplifier analysis. IEEE Trans. on Education **17**(4), 192–198 (Nov. 1974)

Index

C.A. Murillo et al., *Voltage-to-Frequency Converters: CMOS Design and Implementation*, 137
Analog Circuits and Signal Processing, DOI 10.1007/978-1-4614-6237-8,
© Springer Science+Business Media New York 2013